Foreword

Mars 1999 is one of the most important books you will ever read. What it proposes—a manned mission to Mars—is already accepted in many scientific circles as the wave of the future. And on this wave rides an incoming storm of stunning political and economic upheavals.

Most likely among the first areas to be affected will be the precious metals markets. As Dr. O'Leary reveals in this book, we can expect to find vast amounts of precious metals—richer than the richest ores anywhere on earth—on some of the asteroids and, perhaps, on the two moons of Mars.

The closer we get to a manned Mars landing, the greater will grow the realization of those precious metals awaiting us in space. As that growing realization sinks in, more and more buyers and dealers of those precious metals will abandon ordinary bullion, either upgrading to rare precious metal collectibles or leaving the field entirely. But one thing is clear: The closer we get to a manned mission to Mars, the more likely are precious metal bullion prices to plummet.

And a manned Mars mission *is* coming soon. It is imperative we start planning for it now, either with the Russians or without them. For, if we do not, they are preparing to go alone.

We cannot allow that. To do so—to let anyone land on Mars before us—would relegate us to becoming a second-rate power for the rest of our lifetimes and beyond, and would mean the loss of a trillion dollar economic opportunity.

We must not let anyone wrest this opportunity from us. For, whoever gets to Mars and its moons first will have first claim on the vast economic wealth it offers and the megapower that goes with it.

Who knows what other discoveries this venture will unveil? Curiosity has been growing about the strange formations on the Cydonia regions of Mars (as shown in the NASA photographs on pages 38 and 39). If close-up examination ever reveals these to be creations by intelligent beings rather than merely some freak of nature, the ramifications would be staggering. Regardless, the political and economic realities alone of a mission to Mars make this a once-in-a-lifetime opportunity.

NOW is the time to start preparing for it—and to make sure that no one gets there before us.

Alan Shawn Feinstein

February 4, 1989

Alan Shawn Feinstein is a noted financial writer. His newsletters—the Insiders Report *and* The Wealth Maker—*are considered to have the largest readership of any financial newsletters in the country. He is also founder of the World Hunger Center at Brown University.*

MARS 1999

Exclusive Preview of the U.S.–Soviet Manned Mission

Brian O'Leary

Stackpole Books

Published by
STACKPOLE BOOKS
Cameron and Kelker Streets
P.O. Box 1831
Harrisburg, PA 17105

Printed in the United States of America

10   9   8   7   6   5   4   3

Library of Congress Cataloging-in-Publication Data

O'Leary, Brian, 1940–
    Mars 1999.

    Includes index.
    1. Voyages, Imaginary.   2. Space flight to Mars.
I. Title.
TL788.7.O44   1987   919.9'23   87-15481
ISBN 0-8117-0982-5

With Love and Light
To two very special people
born on September 24:
John-Roger, my Teacher,
And Laurene Johnson, who
came into my life well before
the fictional Marla Lee,
for which I am grateful.

Contents

Preface to Second Printing

Much has happened (and not happened) regarding human missions to Mars since I wrote *Mars 1999* two years ago. No president has yet set a dramatic goal as Kennedy did in 1961. The new NASA Office of Exploration, which I briefed in 1987 about the Mars 1999 concept, has set as one of its options a manned mission to Phobos as early as 2003. Interest abroad keeps the flame burning that we can cooperatively send people to the Red Planet before the turn of the millennium. But time is running short for a U.S. commitment to participate.

Soviet space leader Roald Sagdeev has expressed a desire to have the joint manned missions take place in the year 2001. Cosmonaut Georgi Grechko has publicly endorsed this idea: "Let's go to Mars together [near the turn of the century]," he told Americans. "If you won't join us, we'll go alone."

Many experts believe the Russians will go to Mars regardless. Analysts at Commercial Space Technologies, Ltd., report that the Soviets could easily adapt existing hardware (the Mir space station, Proton booster upper stages, Krant and Soyuz modules, and Energia booster) and mount the mission as early as 1994.[1]

In January 1988 I visited the Soviet Union. At the Soviet Space Research

Institute, a technical lecture on the Mars 1999 concept had to be switched from a conference room seating 30 to an auditorium seating 350. The auditorium overflowed. A leading Soviet publisher expressed his interest in printing 100,000 copies of a translation of *Mars 1999*. Last fall, when the Japanese published a translation, I made a major lecture tour to their country.

Will we wake up? Within the NASA, SDI, and other federal bureaucracies, we have massive resources that could be reallocated to Mars 1999, cleaning the environment, developing efficient energy, transportation, food, and health care. I am convinced the American people want this, and President Bush has an unprecedented opportunity to open up our future toward a peaceful, exciting transition into the 21st century.

Brian O'Leary
February 1989

---

[1]S. T. Perera. Paper presented at the 1988 British Interplanetary Society Technical Forum.

Acknowledgments

I have felt inspired writing this book. For me, the action of writing is usually the final stage of thoughts bubbling within for a long period. I write in spurts, and this can often be unsettling to those around me. Thus I am grateful for the loving support of the following individuals during this process:

John Morgan and Bill Haynes, whose comments on the book's contents were both timely and essential; Margaret Jones-Ryan, who supervised the word processing and pulled through at crucial moments, ably assisted by Susan Hauser; Joanne Gabrynowicz, Robert Blunt, Victoria Montgomery, Cynthia Thompson, Diana Judith, Dee Davenport, John-Roger, Whitley Strieber, Christina Baroque, and Laurene Johnson, for their moral support over difficult times; and David Nixon, who compassionately supported the visualization of the Mars 1999 concept.

Many of the illustrations unique to this book were provided by David Nixon's students at the Space Projects Unit of the Southern California Institute of Architecture. Those students include Gary Alzona, Joe Kennedy, Jamie Gregory, Yuchun Lin, Jon Lynch, Chris Miller, Jim Packard, Pat Scarlett, Kristi Skelton, and Iwan Tijonu.

Introduction

Soon we will be sending people to Mars. The consensus building rapidly, both in the United States and in the Soviet Union, indicates that this is so. Planning efforts have begun, and a bewildering variety of mission approaches are emerging.

This book is about one of these approaches. It involves going first to Phobos and Deimos, the tiny potato-shaped moons of Mars. As surprising as it may sound, Phobos and Deimos (known collectively as PhD) are more accessible to Earth, in terms of rocket propellant required, than any other known natural object in the solar system, including our own Moon. Because they have virtually no gravity to overcome, landings and takeoffs are far easier. The PhD moonlets are even easier to get to and from than the twenty-four-hour orbits currently inhabited by most of our communications satellites.

Locked up in their soil, the moons of Mars are likely to contain large quantities of water that can be easily driven out and processed into propel-lants by lightweight solar furnaces. This book will demonstrate that we will need only four missions to PhD, starting with the Mars 1999 project, to create bases on the Moon and on Mars and to develop a self-supporting space economy as early as 2005. We will then no longer need to rely on the expensive launching of fuel and materials from the "deep gravity well" of Earth.

Most of my life has been dedicated to finding ways of creating a human presence in space: to open the option of leaving our home planet, to lift ourselves from our gravity bondage, to create bases on the Moon and on Mars, and to unravel the mystery of life by finding signs of life elsewhere.

That search has led me to the realization that most of the raw materials for any rapid growth of human enterprise in space must come not from Earth but from elsewhere. It is phenomenally expensive to break free of Earth's gravity and launch payloads into space: currently thousands of dollars per kilogram. By contrast, launches from our Moon, or from Phobos and Deimos, could cost *tens* of dollars per kilogram. My col-league Gerard O'Neill was the first to express this in his landmark book, *The High Frontier*. NASA-sponsored engineering studies over the years conclusively confirm O'Neill's ideas.

Most scientists have been looking toward the surface of our Moon as the first step. Here they propose to process raw materials for a growing space enterprise. Chemically bound oxygen composes 40 percent of the lunar soil. When removed, it can be used as the oxidizer component of rocket propellants for launches from the Moon into outer space or trans-ported to Earth orbit for launches of humans and cargoes from there to points beyond. Studies have shown that it is less expensive to deliver lunar oxygen to spacecraft in Earth orbit than to bring the oxygen from the deep gravity well of the Earth. But initial investments in a lunar base would be very high, and lunar fuel delivery buildup in Earth orbit would be slow. Electromagnetic catapults, called mass drivers, on the lunar surface could be able to launch much more material than conventional rockets, but a substantial investment in a lunar base and processing plant and launch system would still have to be made.

This book provides an encouraging alternative, both technically and economically, with the surprising result that we could establish a lunar base and a Mars base much sooner—between the years 2003 and 2005—if we go to PhD first!

Several years ago, I began looking at asteroids as a source of raw materials for economically competitive uses in Earth orbit. Because they have little gravity and some of them occupy orbits close to Earth's, it seemed more fuel-efficient to bring materials to Earth orbit from these asteroids than from the Moon or Earth. In addition, the study of several asteroids revealed that they contain an abundance of free metals, while others contain easily extractable water and carbon compounds. Eventually, billions of dollars would be saved and a self-sufficient industrial economy could be established in space by relying on the economical recovery of nonterrestrial resources.

Over the past decade, however, many of us space futurists were disappointed in the evident lack of positive response from NASA and the White House to the new space exploration ideas and long-term goals. The Apollo days of "going for it" appeared to be over. A more pragmatic approach would be needed to renew their enthusiasm.

We needed to unlock a fresh source of inspiration. We needed to identify an interim key step that would fly on its own merits and at the same time create a space program renaissance as its inevitable byproduct. Over the past several years, my quest has been to identify that step, and I am happy to say that this quest appears to have been successful. The key step we have been seeking is Mars 1999, and the keyhole is PhD.

The concepts described in this book have been published, peer reviewed, and presented at numerous conferences, seminars, and briefings at such places as the NASA Marshall Space Flight Center, Los Alamos National Laboratories, the National Academy of Sciences, the Case for Mars conferences, the Princeton-AIAA conferences on space manufacturing, and the California Space Institute. Colleagues have accepted and expanded on the concepts. The happy result is that there are no showstopping surprise announcements in this book. Mars 1999 and three follow-on missions to PhD appear to be the most economically and technically feasible road to true self-sufficiency beyond Earth.

These ideas are the culmination of a long voyage of the mind, one that has spanned thirty-nine years. I was "initiated" into space in the fall of 1948; I was eight years old. At Harvard College Observatory, where my parents had taken me to an open house, I looked through the telescope at

Mars 1999 Program
Key Events

- THE LATE 1980s—In an unprecedented accord between the U.S. President and Soviet Premier Gorbachev, the two nations announce the goal of cooperative, parallel U.S. and Soviet manned missions to Mars before the turn of the century as part of a nuclear disarmament package.

- JANUARY 24, 1998—The U.S. crew boards the space shuttle *Atlantis* for an orbital rendezvous with the U.S. space station and the Mars interplanetary spacecraft, the *Olive Branch*.

- JANUARY 27, 1998—After a three-day checkout, both the U.S. and USSR Mars spacecraft escape from Earth on a trajectory toward Venus.

- FEBRUARY 6, 1998—First U.S.-USSR crew exchange and misunderstandings.

- APRIL 1998—Resumption of crew exchanges.

- JULY 4, 1998—With a Venus swingby, the two spacecrafts' trajectories are redirected toward Mars, saving time and fuel.

- OCTOBER 31, 1998—Command of the *Olive Branch* changes hands.

- DECEMBER 2, 1998—The U.S. and Soviet spacecraft arrive near Mars and rendezvous with Phobos. Remote-controlled unmanned craft begin the exploration of Mars.

- DECEMBER 5, 1998—The fuel-processing plant on Phobos begins its chores.

- JANUARY 25, 1999—A second U.S. manned spacecraft, *Enterprise,* and its crew escape Earth orbit on a trajectory toward Mars.

- JANUARY 27, 1999—One Soviet cosmonaut from *Mars 1* and one U.S. astronaut from the *Olive Branch* make the sortie in the *Unity* from Phobos to the surface of Mars and return to Phobos.

- JANUARY 31, 1999—Both U.S. and Soviet crews aboard the *Olive Branch* and *Mars 1* launch from Phobos for the journey back to Earth. The fuel-processing plant on Phobos continues to stockpile fuel.

- SEPTEMBER 9, 1999—The *Enterprise* and crew arrive at Phobos and establish a second fuel plant.

- NOVEMBER 9, 1999—The *Olive Branch* and *Mars 1* and crews land on Earth for the completed mission.
- DECEMBER 31, 1999—While Earth residents celebrate the turning of the millennium, the second U.S. crew collects fossils on Mars. Almost complete nuclear disarmament is achieved.
- FEBRUARY 29, 2000—The *Enterprise* crew encounters extraterrestrial intelligence on Mars.
- JANUARY 2001—The *Enterprise* crew, having spent sixteen months on Mars and Phobos, launches Earthward with a payload of 2,500 metric tons of Phobos water. The *Olive Branch* with its crew escapes Earth orbit for its second (and the United States' third) journey to Mars.
- SEPTEMBER 2001—The *Enterprise* arrives at Earth with a big haul of water; the *Olive Branch* arrives at Mars and Phobos to double fuel-processing capacity.
- FEBRUARY 2003—The *Olive Branch* leaves Mars with a second load of Phobos water, and the *Enterprise* is again launched toward Mars on the fourth U.S. manned mission.
- OCTOBER 2003—The *Olive Branch* arrives at Earth with a second big haul of water; the *Enterprise* arrives at Mars.
- 2005—A total of 10,000 metric tons of Phobos water delivered to Earth orbit in four missions supplies enough fuel to launch all the elements of permanent lunar and Mars bases, which are established later that year.
- 2010—Space colonies and settlements begin to spread across the solar system; energy to Earth is supplied by solar power satellites; asteroids are mined for metals.
- 2020—We have established world government, peace, telepathic communication with one another and with extraterrestrial beings; we are in a new age.

Saturn and its rings. The astronomer told me this was a planet bigger than Earth and one billion miles away! I was awestruck.

Soon after, I looked through a telescope at Mars, a beautiful shimmering orange disc with the white speck of a polar cap. I read articles by Wernher von Braun, Willy Ley, and others describing the practicality of space travel to the Moon and Mars before the year 2000. At that point I wanted to become an astronaut.

During school I pursued space and planetary studies before they were accepted in formal curricula. Both my master's and doctoral theses were about Mars, and I was one of about ten doctorate specialists in planetary science in the world (now there are more than a thousand).

In 1967, I was appointed to NASA as a scientist-astronaut. My just-completed Ph.D. thesis dealt with observations of the surface of Mars, and NASA was interested in that research. And so I was picked as an astronaut, the only one in the corps who specialized in Mars. NASA planners were then talking about making trips to Mars in the 1980s, a concept that was reinforced publicly two years later in a speech by Vice President Spiro Agnew. Unfortunately, the nation and the administration were soon distracted by the rigors of the Vietnam war, and the plans evaporated.

Today, as in the days of the Apollo program, the Soviets and Americans are once again on a collision course driven by an extraterrestrial imperative. This time the imperative is Mars. The coming years will be crucial in deciding whether we join our courses or pursue them competitively, as before. This will also be a crucial time to determine how to do it, so the political systems can become educated and take the appropriate action. In the United States, the center of the action is in the presidency. President Reagan has a unique opportunity to leave office with the legacy of a historic decision that has implications transcending even John F. Kennedy's Apollo initiative.

For the past twenty years, Mars has been put on the back burner of NASA planning, but things are changing. During the hiatus, many of us have had an opportunity to develop an entirely new perspective, one that goes way beyond Mars. In this book we shall see that the Mars mission is the linchpin of the forthcoming space enterprise. The political, economic, scientific, and spiritual rationales for beginning this, the largest physical adventure and transformation in recorded human history, are so compelling that we can no longer behave as if they do not exist. We are at an evolutionary point as profound as the time our first human bodies appeared on the face of the earth. The forces of our own growth propel us inexorably onward.

And yet, so many of us appear not quite ready to leap. We are absorbed by the near-term concerns of crisis management, the *Challenger* disaster, protection of vested interests, and conflicting national goals and priorities. It is only recently that long-term planning in the U.S. space program has received new attention. During the intervening years, the grand vision was carried by only a few scientists and space enthusiasts. Most recently, it has been articulated by the President's National Commission on Space.

A human presence in space beyond our Earth is once again being embraced by NASA, and choices made in the next few years in advanced planning will profoundly affect how the U.S. space program will unfold over the next several decades. There are many scenarios. But most of those coming from the advanced-planning community have tended to emphasize infrastructure (see Glossary) rather than a specific goal. There is good reason for this: a broader-based program ensures more continuity and options. Specifics have a way of getting changed. Nobody wants another abrupt halt like that of the Apollo program.

But infrastructure alone may not be enough. The space shuttle and the space station serve the infrastructure but may not be exciting enough. Without a goal like that of Apollo that inspires our collective imagination, we don't go far. Well-intentioned, well-planned steps supporting a growing infrastructure may make engineering sense, but this approach alone will not inspire public support. Moreover, the proposed rate of growth of the infrastructure over decades to eventual self-sufficiency may be too slow to sustain public and private enthusiasm.

Public interest has tended to thrive on observable, nearer results. In addition, the economics of the space program show that money allocated now will go much further to achieve a goal in fifteen years than if it is stretched out over thirty years. Moreover, significant return on investment is best achieved sooner than later. Yet Mars 1999 is not an expensive or impulsive crash program.

A major premise of this book is that for a successful manned Mars program we need both an exciting, recognizable goal *and* the guaranteed infrastructure to follow (cheap fuel in Earth orbit and more trips to Mars), one that grows reasonably rapidly—over years, not decades. I call the type of advanced planning that will be required "hourglass planning." We channel currently diffuse activities to a focal point, as in the Apollo program, but also gear them toward follow-on activities that expand economically after completion of the initial goal. The expansion involves a growing fuel processing plant on Phobos that would drastically reduce the cost of transportation between any two points in space within the inner solar

THREE ADVANCED PLANNING MODELS

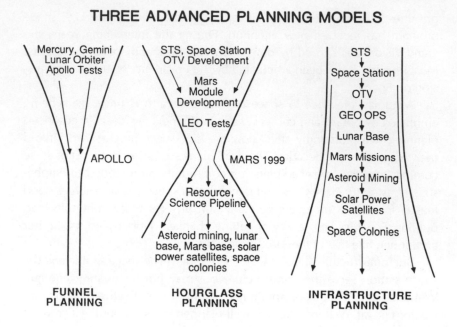

system. In other words, we arrange that a follow-on infrastructure is built into the program.

The Apollo program might be called "funnel planning," one with an inadvertent cutoff after its goal is achieved: there was no follow-on infrastructure. At the other extreme, pure "infrastructure planning" may make rational sense but not acquire the needed political or public support over a reasonable period of time.

This book presents a special blend of rationales, outlooks, institutions, and individuals all pointing to one focus, the constriction point of the hourglass: Mars 1999. The space program six years on the other side of the hourglass midpoint may not be recognizable even to the more sophisticated mission planner. Like the grains of sand seeking to get through the constriction, what happens thereafter is a free fall into a future beyond our wildest dreams.

And yet the simple observable steps along the way are easy to follow and require no esoteric technologies, no real development expense beyond refabricated hardware and some testing in Earth orbit during the 1990s of new ways of doing old things.

Mars 1999 and its three follow-on missions would consume the bulk of NASA's efforts between about 1994 and 2004, a time when the space-station development costs will be winding down. It would be NASA's next major step beyond the space station. But what do we get for it? Permanent habitations on the Moon and Mars; inexpensive fuel and spacecraft re-cycling between Earth orbit, PhD, Mars, the Moon, the asteroids, and other points; and complete commercial dominance of space operations by 2005. Mars 1999 can be achieved well within NASA's performance con-straints and budget schedule, well under 1 percent of the U.S. federal budget. The price for the United States would be even cheaper if the costs were shared by other nations.

Perhaps most important, the Mars imperative provides an opportunity for the superpowers and other nations to come together on common ground. Mars 1999 could become a model for world cooperation and for turning swords into plowshares. It is an idea whose time has finally come.

1

Scenario
Landing in Valles Marineris
January 27, 1999

• *One Soviet cosmonaut and one U.S. astronaut make the sortie from Phobos to the surface of Mars and return to Phobos.*

It was my fifty-ninth birthday. The historic moment we were all waiting for had arrived. Putting on my helmet aboard the *Olive Branch,* about to climb into the *Unity,* I felt a rush of adrenalin. If all went well, the Mars Excursion Module called *Unity* would land Sevastyanov and me in Valles Marineris just three hours later. I was in one of those euphoric states where all of eternity seemed squashed into one moment.

As I turned toward the dark tunnel of the MEM I caught a glimpse of my eyes reflecting off the inside of the helmet faceplate. "The windows to the soul," I said to myself.

Turning into the bright light of the cabin, I knew that ten minutes from now, 150 million kilometers away, about six billion earthly eyes would see mine. Ten minutes from now I would be on the couch of the MEM that would soon take Sevastyanov and me to the surface of Mars.

Yet in my heightened awareness, I saw and felt those eyes at the instant I turned back into the tunnel, and sensed a unity with all humankind. The speed of light is *not* the speed limit of consciousness, I reflected. Einstein was only talking about electromagnetic waves, light and radio communications that created the illusion of separateness between the *Olive Branch* and Earth. Actually I felt grateful for the time delays of minutes to Earth; it made us more autonomous.

Bulky suit and all, I floated through the hub module docking adaptor into the cone-shaped MEM. Now I was on the couch, sitting in the right-hand seat. Sevastyanov was already there with his space suit on, going through his checklist. I glanced out the window at an awesome sight I never seemed to tire of. In one direction the huge orange disc of Mars covered a forty-three-degree diameter chunk of the sky. Off to the side, the potato-shaped rocky moon, Phobos, twenty kilometers wide, covered an even larger slice of sky. Even the brilliant cinematography in *2001*, *2010*, *Star Wars*, and *Star Trek* couldn't match the view from the *Olive Branch*.

I reflected on my more meditative moments in weightlessness inside the hub module during the year's journey to Mars. "There is something lacking in modern life," the visionary physicist Freeman Dyson had said. "I'd like to go to Mars or some other place simply to find the peace and quiet that used to be [on Earth]."

I had found that, at least at times.

Much of the year-long journey to Mars had not been easy. I'd had some anxious moments, like when the external tank and PhD payloads started tumbling just after leaving Earth orbit, or when the experiment module of the *Olive Branch* heated up to 120 degrees during the Venus swingby and knocked out some of our interplanetary astronomy experiments.

And then there was the first crew exchange with the Soviets. They had offered us vodka to celebrate, and Commander Busby sternly declined. That put a damper on our togetherness, a very bad start. For two months, the crew exchanges between the *Olive Branch* and the Soviet interplanetary spacecraft counterpart, *Mars 1*, were cancelled. I recall looking out the window at the Soviet craft, barely 100 meters away in the cold vastness of interplanetary space. No Earth or Moon stood as references; they had become mere dots in the sky, like Mars, Venus, and the stars. What a chilly, lonely feeling, viewing the only other vestige of human civilization, so near and yet so far.

John Busby, the seasoned fighter jock, man with the right stuff, was unquestionably good in the cockpit. But as leader of five strong individuals on a twenty-two-month voyage to and from Mars, he was out of his element. He knew it, too, and so implicitly gave most of the leadership over to Marla Lee, the *Olive Branch* copilot who did have the right stuff for this kind of trip. Busby was now along for the ride, but he still gave the impression he was in command, a fiction we played along with . . .

Until it was time to land on Mars. Sevastyanov did not want to do it with Busby, and I can't say I blame him. Lee had asked that I take Busby's place, and so here I was in the MEM—fifty-nine years old and certainly no pilot. I was a planetary scientist and senior engineer on the mission. As part of our training I had learned to fly the MEM and deploy Martian experiments, but I never thought I'd actually be doing it. Sevastyanov was an excellent pilot, thank God.

T minus fifteen minutes and counting. The launch window for the landing in the central trough of Valles Marineris was very short—forty-five seconds. If we missed it, we'd have to wait another seven hours and thirty-nine minutes, when we and Phobos were once again in the right orbital position for retrofire to the landing site.

T minus ten minutes. We undocked from the *Olive Branch* and saw it slowly receding. With an unobstructed view, I could now see the long shadows on the evening terminator of Mars directly below us. Soon we would be firing our retrorockets to take us on the languishing ellipse toward the surface of Mars, landing on the opposite side where it was almost morning.

T minus two minutes. Most of our descent would be in solar eclipse, in darkness. We would have to trust our tracking data until minutes before landing. If we were the slightest bit off course, we might have to perform some tricky maneuvers to land in the right spot inside the Grand Canyon of Mars.

T minus sixty seconds, and all was go. We were now in an automatic firing sequence. I recalled the Apollo 11 first lunar landing thirty years ago, when Neil Armstrong and Buzz Aldrin touched down many kilometers off target because gravity anomalies called mascons (mass concentrations) threw them off course. They had landed by the seat of their pants with little fuel left. Pretty gutsy, I thought. The state of the art had advanced a lot, but still things were controlled by small computers on the MEM and *Olive Branch,* not from faraway Earth. We were on our own. Would Mars cons throw us off?

T minus twenty. My pulse quickened and I sensed a lump in my throat. Ten, nine, eight, seven, six, five, four, three, two, one, *fire.*

I felt a pressure pulling me back to the couch. Even from inside the space suit the engines sounded loud. Little had been done for life-support systems on the MEM. The trip was to be a nine-hour sprint—2.1 hours down, a mere 4.7 hours on the surface, and 2.1 hours back to rendezvous with the *Olive Branch.*

We didn't even have cabin pressure. It was like one long EVA.* The ascent would have to take place at the right moment, or we would have only one more chance to come back, ten hours later after Martian nightfall.

T plus fifty seconds; deorbit engine cutoff. Spinning slowly on its tether, the *Olive Branch* now was barely perceptible and receding rapidly above and ahead of us. Next to it was *Mars 1.* Both spacecraft were dwarfed by Phobos, which was also getting smaller by the second.

Our mission was primarily political. Pains had been taken to bring along an aluminum platform for the bottom of the MEM ladder, big enough to fit two of us, so the first human steps on Mars could be taken simultaneously by a Russian and an American. Buzz Aldrin would have probably liked to have that on the Moon, I was thinking.

It was also the first mission to use nonterrestrial materials for fuel. In spite of the problems with Busby, the PhD water and fuel processing plant worked almost perfectly from the start. We were using close to forty metric tons of liquid oxygen and liquid hydrogen to deorbit, land on Mars, take off, and rendezvous with the *Olive Branch.* Eldon Steinmuller from Germany was the right man for the job. He was a brilliant engineer who had spent the last ten years of his life tirelessly designing, troubleshooting, and training himself to manage the PhD plant, which would become the economic doorway to a space renaissance only five years later. Ph is for Phobos, D for Deimos, the two moons of Mars.

T plus fifteen minutes. As the Sun disappeared and we entered eclipse, I noticed the Martian horizon was less curved. We were getting closer. I looked over at Sevastyanov and we exchanged grins. He knew little English and I knew little Russian, but for twenty years we had had a rapport of the heart that made us a good team. And he really knew the MEM well, even though it was built by Grumman, an American company. He had spent a lot of time in the United States learning how to fly it. Most of the mission would be automatic except for the final phases of landing;

---

*See Glossary.

boulder and crater avoidance could still be best done through the eyes of a human. I was to read out the instruments while Sevastyanov landed the MEM. We were each in communication with our control centers in the *Olive Branch* and *Mars 1*, which had a complicated command coordination procedure that was almost entirely automated.

T plus one hour, fifteen minutes, more than halfway through the descent. Here we were, one cosmonaut and one astronaut, temporarily cut off from the Sun and Earth, no signs of humanity, not even the visual comfort of the *Olive Branch*, the *Mars 1* or the Phobos base. All of us were in darkness, on the night side of a mysterious planet heretofore untouched by human hands or feet. And the rovers and orbiters were telling us Mars once harbored life forms.

On-board computers were telling us to trust them, that we were soon going to be breaking into daylight and would be entering the atmosphere over a rapidly moving red terrain. During the final descent, we would be dropping into a canyon four times as deep as the Grand Canyon and as long as North America is wide.

It was all unreal, I thought. But was the Earth any more real? I felt that I was suspended in a twilight zone caught between life and death.

T plus one hour, thirty-two minutes. A real twilight zone began to materialize outside the window. Our cone-shaped and somewhat elongated spacecraft, resembling an Apollo command and short service module, was moving blunt end forward toward the light band on the eastern horizon, whose curvature was now barely perceptible. We were only 120 kilometers above the Martian surface traveling at four kilometers per second, slowly dropping and about to enter the atmosphere. The sunrise band became a beautiful red and yellow.

T plus one hour, thirty-four minutes. Sevastyanov was busily going through his checklist, speaking in Russian to his command center on *Mars 1*. No glitches, a smooth flight.

T plus one hour, fifty minutes. We were sixteen minutes from Mars touchdown. Suddenly red mountaintops reflecting early-morning light began to appear below. These were the giant extinct volcanoes of the Tharsis Ridge, an extensive high plateau near the equator to the west of Valles Marineris. Up toward the north, I could see the vast red hood of Olympus Mons, the highest mountain on Mars. It stands three times the height of Mt. Everest and is as big as the state of Colorado. On the lee side of the giant Olympus, I could see the white plume of a cloud. The sight was

Just before landing, the crew of the *Unity* views the Martian surface. *Photo taken by the Viking orbiter in 1976.*

awesome, much more real than the more static, distant views from the telescope on Phobos.

T minus thirteen minutes from touchdown. At first contact with the sparse outer fringes of the atmosphere, we felt a gentle pressing back on the couch. Entry began. As the g's* began to build, we could see the landscape whipping along below. Even Disney couldn't have invented

---

*See Glossary.

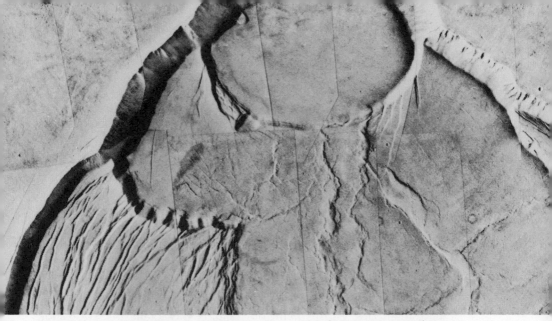

The summit crater of the giant volcano Olympus Mons, which the *Unity* crew sees sixteen minutes from touchdown on Mars. *Mosaic of thirteen photos taken by the Viking orbiter in 1977.*

such a sight! We had viewed a videotaped simulated sequence of the approach, but it had not prepared me for the real thing—mountains, mesas, craters, and then, over the lip of Valles Marineris, which we would be following for another several hundred kilometers, there was the most incredible sight of all: a redder-than-red Grand Canyon.

T minus ten minutes. The maximum g-loading pushed us back into our couches, and my job of reading out status checks began in earnest. Soon I would not be able to afford the leisure of looking out the window. Except once, at T minus five minutes. Our descent was almost vertical now, and suddenly out the window appeared towering red cliffs. In the morning light they looked like hundreds of Zion canyons, indescribably deep with mesas, walls, landslides, horizontal layering, and tributary canyons in all directions. An incredible view! We were now over the lip and in the canyon, still dropping and still three minutes from touchdown. At this point the canyon was as deep as Mt. Everest is high, ten kilometers. Communicating with the control center on the *Olive Branch,* I forgot my professional manner.

"Wow! What a beautiful view! Red mountains and cliffs everywhere and incredible striations! It's so deep. The sky is purple. I've never seen anything like this before. There was *certainly* water here, and I see a layer of rock over there that's blue-green!"

"We copy," said Lee on board the *Olive Branch*. "Your checklist targeting status, please. We have you on a 200-meter downrange overshoot. Some boulders there and a little close to the Y-Mesa. Suggest five-second 50 percent throttle burn to initiate in eight seconds."

"Roger." I looked over to Sevastyanov, who was now looking out the window toward the target zone. I punched in RCS 5 at 50, which appeared on a screen in front of the two of us. Soon the same command went from *Mars 1* to Sevastyanov, who fired the engines. We jolted a little as we descended farther, now enveloped within an enormous canyon. "On target now," said Lee, "T minus seven-zero seconds. Keep your eyes on the panel, we want you back here."

I began reading the altimeter. "T minus thirty, 315 meters."

Sevastyanov moved his joy stick, causing jolts to the left. Then forward, then back. At this point our landing was totally in Sevastyanov's hands, as he eyeballed the smoothest site.

"Eighty meters . . . fifty meters . . . thirty meters . . . twenty meters . . . fifteen meters . . . eight, seven, six, five, four, three, two, one, touchdown." Jolt, rebound, settle at a slight tilt.

Sudden silence. It seemed an eternity. Then Sevastyanov and I spoke simultaneously. In Russian he said to *Mars 1*, "The *Unity* has landed safely." To the *Olive Branch* I said, "The *Unity* is safe and sound in Valles Marineris base." The world would hear this ten minutes later.

Like two awkward alien bears, Sevastyanov and I hugged, as well as we could in our space suits in a cramped cabin. We smiled warmly at each other, and I yelled, "We did it!"

"Yes, comrade, we did it!" He let out a loud howl that pierced my ears over the intercom. I was tempted to let out a howl too, but then I remembered the television camera trained on our faces, three billion pairs of eyes looking at our every expression.

I looked out the window and spoke into the intercom. "The Marscape everywhere is so incredibly red, like an eerie luminescent paint coating on every rock, undulation, and sand dune and all over the cliffs and mesas in the background. The cliffs are not as steep as I expected, but I can get a feeling of their tremendous height. Even the sky is red! It's as if some crazed environmental designer came upon this planet and sold the Martians a total look—only one kind of paint for everything. Trillions of cans of iron oxide. Good God, this is really something!"

"Roger, *Unity*," said Lee. "Time for your pre-EVA checklist."

Again, for a few seconds, my excitement had overwhelmed my sense of responsibility. But there was no time to waste. All within four and one-half hours, between 8:00 A.M. and 12:30 P.M. Mars time, we would have as busy a time line as ever conceived on this mission. We would be checking ascent stage systems in case of the need for an emergency takeoff at any time. We would be collecting contingency samples. We would be doing a busy EVA (extravehicular activity: NASA-ese for a Mars walk). We would hold a ceremony and deploy a large number of scientific instruments and materials processing plants. We would be walking, looking for signs of life, taking pictures, and collecting samples.

Why such a short trip? This had been debated for years. The geologists wanted us to stay for a few days, but the engineers and space agencies of both the United States and USSR said no. There were just too many risks, too much expense for the first landing. A mission of just nine hours needed no life-support system for the cabin and no Martian stays during the cold, dark nighttime. It also meant that co-orbiting Phobos, *Olive Branch* and *Mars 1* would be in communication and tracking line-of-view for the entire mission, in case the Martian relay communications system wasn't up yet. And the second Mars landing, planned for December 1999, would bring along a Mars hab module that would house a crew for more than a year.

The landing site had required much less debate. Valles Marineris (Mariner Valley) was first discovered by and named for the unmanned U.S. Mariner 9 spacecraft in 1971. This enormous rift canyon runs along the equator and shows abundant evidence that subsurface water once poured in along channels. Landslides, horizontal layering, volcanic vents, and channel outflows into the canyon suggest that parts of the valley floor were once a lake that contained water covered with ice, similar to lakes in the dry Antarctic valleys. Would we find fossils there? Valles Marineris was a prime site for exploration.

Also, by examining the stratification along canyon walls, we would understand the nature of what lies beneath the Martian surface down to ten kilometers. Busby had been cross-trained in geology, and Valles Marineris was his favorite subject. But he was in sick bay with a fever and high blood pressure, burned out and probably envious, maybe furious at us and at himself for his personal decline since the mission began.

In rare agreement with the scientists, engineers favored the Valles Marineris site because it was on the equator. The amount of rocket fuel needed to go from Phobos to the Martian surface and back again was far less if the landing spot was nearest the equator. Phobos's orbit lies along the Martian equator and no costly orbital plane changes were needed. Still, we wanted to find a safe, flat, relatively unrocky landing target that would also be geologically interesting. Marineris base was chosen for that reason, and also because it lay near the mouth of a tributary outflow channel. Diehard engineers who pushed for a bland, flat plains area were overruled.

Another contingent outside the program wanted us to land in Cydonia near the famous "face" on Mars. But because of its latitude—41 degrees— we could never have gotten there from Phobos on the *Unity*; there simply wasn't enough fuel. One of our unmanned rovers had landed on the west side of the face at the beginning of our stay at Mars. Just two weeks before, it had photographed and chemically analyzed mysterious metallic flakes that suggested some form of intelligence might have been there. This had been quite a month!

T plus fifteen minutes. The historic moment was rapidly approaching and I had to go to the bathroom! Fortunately our space suits were equipped with urine receptacles.

Sevastyanov was first to leave the *Unity*. The door opened to a sudden breeze that ruffled some of my papers. Even though we were at one of the lowest points of Mars and on the equator, the ambient pressure was less than one-hundredth that of our atmosphere; the temperature was climbing but still at minus 40 degrees, both Celsius and Fahrenheit. Water was once present on Mars, but now the atmospheric pressure was too low for it to condense and pool, even in the depths of this valley.

At the time of our landing Mars was near aphelion, the farthest point in its orbit around the Sun. The good news about aphelion was there were no big dust storms raging on Mars during our visit. The bad news was it was colder in the weaker sunlight, and our solar power budget was tight.

I unhooked my life-support hose, went on internal, and climbed out onto the ladder. Sevastyanov waited on the platform a mere half-meter from the Martian surface. One-third g, I felt, was just about perfect. This was the level we had selected for the *Olive Branch* after years of Earth orbital testing, enough to prevent bone decalcification and readaptation to higher g-levels yet low enough so that the tether and tunnel joining the hab modules to the hub module was a reasonably short sixty meters. We

The *Unity* crew plants American and Russian flags on the Martian surface during a brief ceremony. *Artist's conception by Paul Hudson, Orbital Sciences Corporation.*

enjoyed weighing one-third as much as we did on Earth, heavy enough to restrain ourselves moving around and light enough to bounce around.

I joined Sevastyanov on the platform. We looked at each other and then, touching EVA gloves, we stepped onto the red Martian sand at the same moment, 22:35 GMT, January 27, 1999. We walked a few steps and turned around toward the MEM television camera. Simultaneously in Russian and English, still joining hands, Sevastyanov and I read this statement we had created on the voyage:

"We bring you our joint personal message from the surface of Mars. While we represent our respective governments of the USA and USSR, we are really representatives of humanity, in this the greatest of human adventures. We are a small symbol, a temporary focal point representing the need for the entire Earth to unify, regardless of historical differences in philosophical points of view and emotional attitudes. Let our unity as two humans visiting Mars carry a message to the entire species back home on the planet Earth, so that on the occasion of our return in November we shall be able to join you in celebrating the arrival of the year 2000. Let us finally complete our journey to nuclear disarmament and declare a lasting peace and loving cooperation. We are committed to transforming bloody old swords to shining new plowshares as we enter the third millennium A.D. of the world calendar. Let us open up the resources of outer space to a human renaissance of which our presence here is yet another step. Let us join in exploring the cosmos, in which we can live and evolve beyond our Earth beginnings. Let us carry and share our life and love everywhere."

2

Why Mars?

Is this science fiction? Some of the characters, names, specific times, and sites are, but the basic facts are not. We know enough now to successfully complete Mars 1999, and in a way not unlike the scenario in chapter 1. If anything, the truth that emerges will be stranger than the fiction projected.

Mars beckons humanity. Why? Because it appears to be the next logical extension of "home" for a growing human population. Photographs taken from Mars orbit aboard unmanned Mariner and Viking spacecraft reveal dried-up river beds, huge volcanoes, and grand canyons that "cry out for exploration," in Carl Sagan's words. Mars, of all the other planets near us, looks most familiar and hospitable.

Earlier epochs on Mars appear to have been earthlike, including an abundance of water on the surface. This suggests the tantalizing possibility of finding evidence of life, perhaps in the form of fossils. It has even been suggested that certain features on Mars resembling a face and huge pyramids could have been created by an intelligent civilization. Confirmation of that would revolutionize our understanding of our origins and role in the cosmos.

We have an opportunity as never before to go there and check all this out. How and when, that is the question. This book describes a means for beginning the grand adventure in 1999 in a manner that is as inexpensive as alternative schemes and yet would yield quantum-leap dividends in the scientific exploration of Mars and economical recovery of water and fuel from Phobos and Deimos.

The scientific community knows that the two Mars moons, Phobos and Deimos, are more accessible to Earth more often than any other natural objects in the solar system. That these Martian satellites are almost certain to contain water that could be used as propellant for subsequent missions and for establishing bases on the Moon and Mars is only part of the story. As these economically logical activities begin, looming in the sky above the PhD water and fuel processing plant is a huge orange balloon called Mars. Mars is the bonus.

And it's a bonanza for scientists, who have often felt left out of the space mainstream. After a golden age of solar system exploration with unmanned spacecraft mostly during the 1970s, we are in a hiatus. The solar system has been put on hold, and many planetary scientists are disturbed, resentful, and defensive. One of them, James van Allen, the man who in 1959 discovered Earth's radiation belts later named after him, has even proposed that the manned space program be scrubbed. He believes we should concentrate on expanding on our success in sending instrumented probes throughout the solar system.

Most other planetary scientists are less extreme, yet they reflect the same sense of conservatism that currently pervades NASA. They are talking about a phased program of precursor missions to Mars before sending people there. We hear arguments of manned versus unmanned space flights, of humans versus machines. The debate dissolves when we consider the Mars 1999 scenario and its follow-on missions.

While we see disagreement on approaches to space exploration, there is an emerging consensus on where we would like to send people—Mars. In spite of their different political and philosophical perspectives, Carl Sagan (astronomer and host of "Cosmos"), Buzz Aldrin (Apollo 11 lunar module pilot), Harrison ("Jack") Schmitt (Apollo 17 lunar module pilot and former U.S. Senator), Bruce Murray (former director of the NASA Jet Propulsion Laboratory), Senator Spark Matsunaga, Congressman George E. Brown, 80 to 90 percent of those responding to public opinion polls, various Soviet groups, the "Mars Underground" (about 100 scientists dedicated to sending people to Mars), popular space groups such as the 100,000-member Planetary Society, the National Space Society (formerly the L-5 Society and the National Space Institute), the *New York Times,* the *Los Angeles Times,* and the President's National Commission on Space . . . all favor a human trip to Mars.

Even NASA is showing renewed interest. The NASA Marshall Space Flight Center in Huntsville, Alabama, has recently doubled and redoubled its budget on Mars manned mission and system studies. And James Fletcher, NASA administrator, has appointed a committee headed by astronaut Sally Ride that has placed the human Mars mission on its list of goals. The bandwagon has begun.

My vision is that we will create on Mars an economic infrastructure in a setting where there is a lot to go around for everybody—scientists, engineers, philosophers, artists, individuals, governments, all of us. But first let us look at the physical characteristics of this world we are turning our technology toward.

By earthly standards, Mars is a barren and arid planet with a surface atmospheric pressure one two-hundredth that on Earth and temperatures well below zero most of the time. Liquid water cannot exist in a free state on Mars; it would boil away or freeze. The thin atmosphere is made almost entirely of carbon dioxide with traces of water vapor. Ancient impact craters over much of the surface suggest an unaltered (unexciting) geological and little if any biological evolution. Life-detection experiments aboard the two NASA Viking landers in 1976 showed ambiguous results and no positive signs of life.

The Mariner 4 mission to Mars in 1965 took twenty-two television prints of 1 percent of the surface—by hindsight, an uninteresting 1 percent. The prints showed craters and no signs of life. The scientists' conclusion was the planet was dead . . . very disappointing.

Astronomers have observed Mars for centuries. It is one-half the size of Earth, has a day slightly longer than Earth's (twenty-four hours, thirty-seven minutes), and an axis tilt giving rise to seasonal variations similar to those we have on Earth. And yet Mars is the next planet out from the Sun, with an average distance 52 percent greater than Earth's. Its orbit is more eccentric than Earth's, moving at variable distances 39 to 62 percent greater. The surface gravity on Mars is one-third that on Earth. Mars is by far the solar

A boulder-strewn field of red rocks reaches to the horizon nearly two miles from Viking 2 on Mars' Utopian Plain. The circular structure at the top is the high-gain antenna, pointed toward Earth. Viking 2 landed September 3, 1976, some 4,600 miles from its twin, Viking 1, which touched down on July 20.

system's prime candidate for extraterrestrial life, past, present, and future.

The Mariner and Viking probes opened our eyes to some extraordinary sights: Valles Marineris, Olympus Mons, the water channels, the polar caps and hood, the dust storms, the pyramids, and the face. The Viking lander discovered a peculiar chemistry that caused one of the life-detection experiments to produce unexpected readings. Scientists began to question whether those experiments were appropriate for the peculiar environment of Mars.

The notion that Mars may once have been teeming with life is rampant in mythology, extending to this very day. Myths may not be physically true, but they often tell stories that lead to a later reality. We have been dancing with Mars myths for quite some time. Early in this century, the wealthy Sir

Percival Lowell dedicated his life to creating an observatory in Flagstaff, Arizona. He believed there were canals on Mars that could carry scarce water from the polar caps to a civilization trying to survive on an arid planet, and he hoped to observe them.

With Lowell's maverick (and later disproved) hypothesis, the quest for life on Mars began. During my college and university years, astronomers in France and the United States observed a wave of darkening of Martian features that advanced southward from the receding polar cap each spring. They suggested it was caused by vegetation, but it was moving in the wrong direction: the darkening wave should have gone from the equator to the pole. It turned out to be the seasonal deposition of dust.

Mars appeared through all earthbound telescopes as a vague shimmering disc, only occasionally sharp enough to reveal variations in the bright-

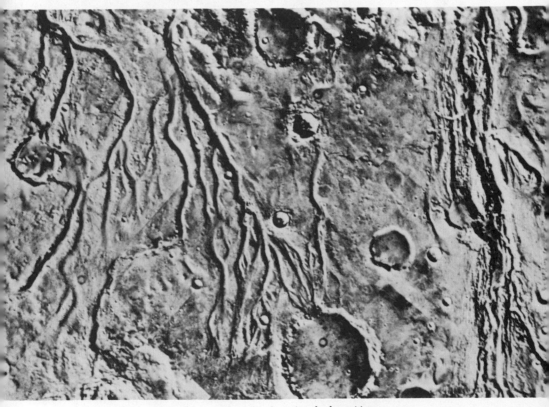

Viking orbiter photograph of dried-up river beds on Mars.

Unusual features in the Cydonia region include a mesa that looks like a human face about one-mile wide (*upper right*) and pyramids (*left*). Circle in the center is a blemish in the Viking camera.

ness of features and the seasonally advancing and receding polar caps. Lacking sophisticated resolution, astronomers before the Space Age resorted to spectroscopy, the science of dividing light into its component colors. If the spectral resolution is high enough and if the surface or atmosphere of Mars is cooperative enough, we can understand what is happening on Mars chemically. Before spacecraft ever went to Mars, spectroscopy told us the atmosphere was made up mostly of carbon dioxide with a trace of water vapor. It also told us that the surface was coated with iron oxides.

During the ten years since the successful completion of the Mars Viking mission, no aspect of Mars study has received more attention and speculation than orbiter imagery of a curious two-kilometer-wide mesa that, to most observers, resembles a human face. Thus far, four books and a score of articles have focused on this and other unusual Martian features in the Cydonia region. Strongly conflicting views have emerged, ranging from the extreme—that intelligent intervention must have created or modified these objects—to the more traditional belief that the objects are completely natural.

Most planetary scientists hold the latter view. Geologists point out that the Cydonia region contains thousands of mesas, some of which are bound to appear facelike. In 1976 Viking scientists dismissed the face as a trick of lighting and shadow, but later studies showed that the three-dimensional structure of the face is real and that the structure is facelike at two different solar illuminations, 10 degrees and 27 degrees above the western horizon.

Are these completely natural formations, or were they carved out by a civilization? Science writer Richard Hoagland and others have argued the case for intelligent intervention. They base their hypothesis on the apparently improbable statistics of the spacings and directions between key objects and, of course, their curious forms. My own opinion is that no uniquely intelligent intervention can be persuasively argued on a statistical or geometric basis at this time; the observed alignments and spacings are not sufficiently distinct and evaluation criteria are too vague. To test hypoth-

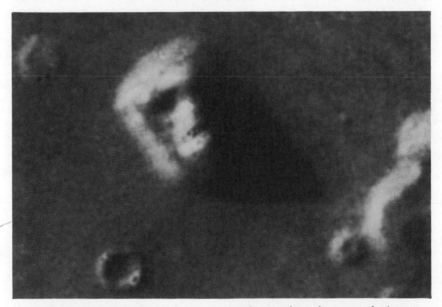

Close-up photo of the "face" with the Sun 10 degrees above the western horizon.

eses based on geometry, we must await higher-resolution imaging from the Mars Observer unmanned orbiter planned for 1992 and other future missions.

A second approach is to use state-of-the-art image processing algorithms to extract as much data as possible from the Viking images. For an organization of scientists we have called the Mars Anomalies Research Society (MARS), Mark Carlotto of the Analytical Sciences Corporation has produced a number of "cleaned-up" images of the face and a three-dimensional map that resolves any illusions of lighting and shadow. When viewed from above, the three-dimensional structure of the face confirms facelike contours and a considerable degree of symmetry, although not complete symmetry. One could argue that a hypothetical civilization could have carved the face but incomplete construction or subsequent geological processes have altered its symmetry. One result we are reasonably sure of is that if intelligence did create the face, it was meant to be viewed from above and not from the ground.

Some scientists argue that intelligent intervention could not have done anything as bizarre as creating a humanoid face on Mars, as this would violate our operating paradigms in the Search for Extraterrestrial Intelligence (SETI; see Glossary). I believe such a view is overly narrow and that all reasonable inquiries into possible manifestations of extraterrestrial intelligence are worth pursuing. There will always be skepticism among some scientists when a new SETI hypothesis is formed, yet an open mind is essential. I feel no less comfortable about discovering a face on Mars than I would about receiving a radio signal from another solar system or reading a credible UFO report. Whatever the form, the discovery of extraterrestrial intelligence will be "improbable" and surprising, and will transcend provincial terrestrial thinking. I am urging that high-resolution imagery of the face and surrounding region at a variety of lighting conditions be performed as a high priority on the 1992 Mars Observer mission.

Mars is a world awaiting discovery. "Consider," writes Carl Sagan in the December 8, 1986 issue of *Aviation Week and Space Technology,* "roving, microbe-free robots . . . wandering to view close up some of that profusion of Martian wonders. Television images of new terrain and new wonders could grace our home television sets every day for more than a year. The surface area of Mars is almost exactly equal to the land area of the Earth: it provides an ample arena for a new age of exploration."

"Planet Mars has always been alluring to humans," said Bruce Murray, former director of the Jet Propulsion Laboratory. "The other planets are far less attractive as potential destinations. . . . Mars is not only survivable, but in the long run habitable."

The question is not whether we will go to Mars. We will go. The question is when, and then how and who. This book provides a plausible answer to these questions. The scientific bonanza of knowledge offered by the Mars 1999 mission and the three missions to follow exceeds anything imagined by the mission planners who contemplate a slow buildup of exploration. Their agenda includes an Earth-controlled automated rover and sample return, some more unmanned visits, perhaps a manned flyby or Mars orbiters scouting out for bases, a manned landing in one or two sites, and eventual buildup of a Mars base in the third decade of the next century.

This last step is a goal stated in the report by the President's National Commission on Space. NASA planners had put this even further in the future, if ever. Yet, as we shall see, economics alone dictate building the Mars base and lunar base rapidly and concurrently between 2003 and 2005. Meanwhile a dozen or more Martian sites could be explored by unmanned vehicles. The Mars 1999 scenario may sound bold; but it is actually cheaper if done in the earlier time frame with PhD leverage.

The rate of flow of scientific data from unmanned rovers and sample returns from several Martian sites during Mars 1999 and subsequent missions, controlled initially by astronauts near PhD, will be in the millions above that which could be achieved by similar systems directed from Earth. We can select riskier landing and traverse sites because the control feedback will be a fraction of a light second away in a Mars orbit, not ten light minutes away from Earth. In the Earth-controlled scenario, a Mars rover could fall over a precipice or pile into a boulder before we knew about it on Earth and were able to take corrective measures. And we can screen returned Mars samples for contamination on the spot at PhD before giving the go-ahead to land people on Mars.

These considerations lead to a multi-site Martian scientific investigation that staggers the imagination. Precursor missions are really not necessary; they offer intrinsically little compared to what is possible in twelve to fifteen years. Mariner, Viking, the Soviet Phobos mission planned for 1988, and Mars Observer 1992 will be enough!

Planetary scientists can become blinded by their own conservatism.

There's good reason for that: they have, in effect, been left out. Their prefer-
ence is to quietly ease stretched-out, lower-budget programs into mission
planning. Mars 1999 is not easing into it, yet ironically it is the cheaper way
of going about things. The Soviets have given indications that they already
understand this; the United States now needs to lay claim to what is essen-
tially its own idea.

3

Scenario

The Grand Accord
Late 1980s

• *In an unprecedented accord between the U.S. President and Soviet Premier Gorbachev, the two nations announce the goal of cooperative, parallel U.S. and Soviet manned missions to Mars before the turn of the century as part of a nuclear disarmament package.*

How is it I happened to be on Mars in 1999? Twelve years earlier I would never have dreamed it. Come to think of it, twelve years was all it took to create the first lunar landing from virtually no space program at all.

I recall talking and writing about space in high school in 1957 and having some of my teachers react in disbelief. That fall *Sputnik* went up and twelve short years later humans landed on the Moon. Would the man on the street in early 1957 have believed people could walk on the Moon by 1969? Even the visionaries would have had trouble believing that. By the same token, who in 1987 could have believed we would have gotten to Mars by 1999?

Actually, the Mars scenario was more believable than the lunar scenario, because the political and technical stages were already set. The lunar landing concept was only a dream until eight years before it happened, when President John F. Kennedy dramatically announced the goal to Congress on May 25, 1961. Chopping the air with his hand, he said,

"No single space project in this period will be more impressive to man-kind or more important for the long-range exploration of space." The energy behind his speech was extraordinary and the vision was clear. Then there was leadership and the clear statement of a heady goal.

The Martian analogue to Kennedy's speech began with public Mars rallies held during the U.S.–Soviet summit in Geneva on August 6 to 8, 1988, culminating in an agreement on August 8. Eight-eight-eighty-eight. The timing could not have been more cosmically elegant, even though the national leaders may not have been consciously aware of it when the dates were set, for the number eight is also the Mobius loop that signifies infinity. Our physical frontier toward infinity is space, and nothing would release the energy and move us closer to infinity than a human trip to Mars, a voyage one thousand times the length of the Apollo trips to the Moon.

Millions of people celebrated the vision of peace that weekend in events organized long before. In Central Park, entertainment superstars offered a free concert called "Make the Garden Grow All Over the World." Concurrently in Geneva that weekend, an independent peace conference attracted peace ambassadors who walked (where possible) from every nation on the planet to celebrate global harmony. Meanwhile the super-power leaders met in a small villa outside town amid rumors of a break-through in arms negotiations and talk about Mars.

The proposal on the table was the SALT III treaty for phased reduction of nuclear weapons to zero by the turn of the century and tight limitations on research and development of spare weapons on both sides, including the U.S. Strategic Defense Initiative (SDI, or "Star Wars"), to be phased out along with the nuclear cutbacks. France, Great Britain, China, India, Is-rael, and other nuclear powers were concurrently meeting with U.S. and Soviet representatives to phase out their weapons, too. On-site inspection everywhere would enforce the agreement.

Whether such an agreement would stick was another question. It appeared that the Soviets were straining their economy to meet the U.S. SDI challenge, and so felt motivated to propose arms reduction. It was not out of altruism so much as practicality. They didn't want to go bankrupt.

The Mars proposal was certainly a trigger to the arms-reduction agree-ment. The President wanted to carve a niche in history with a lasting and positive program, like Kennedy's Apollo goal. Signing the disarmament treaty removed a negative focus but he also wanted a positive focus. Politically, the President suffered from the embarrassment of recent mistakes. The Mars program, he hoped, would divert public attention away from them, just as Kennedy had used Apollo to shift the focus from his Cuban Bay of Pigs fiasco.

In the election year of 1988 and during the following year of the new administration, it was clear to perceptive leaders of both parties that the government needed to shift jobs over to civilian R&D efforts in energy, the environment, transportation, pure science . . . and space. All these programs provided for shifts toward commercial takeover in ten to twenty years.

Going against the advice of his top aides, the President favored a dramatic package that would culminate in a series of carefully thought-out cooperative missions to Mars. Some key outside people had been able to convince him that the Mars program would not significantly increase NASA's budget, eventually would even pay for itself, that it would be international in scope, that the options of cooperating or competing with the Soviets would remain open right up to the time of the mission.

The President also proposed that the bicentennial of the American Constitution could not be better celebrated than by convening the countries of the world in a process similar to what happened in America 200 years ago. As in the time of the first constitutional convention, the process might take several years, but it could lead to a democratic world government. The role model for making all this happen would be cooperative space exploration, and the first big project would be Mars. The historical fit was perfect.

The energy for these big changes was missing before 1988 and 1989, particularly among the NASA people, who were still digging out of the storm of *Challenger.* But just like Apollo, the initiative had to come from the outside and from the top. In a speech before Congress, the President directed the federal government to conduct studies on the best ways of creating a Mars mission by the end of the century. Could we pull it off, he asked? Let's find out, he ordered NASA.

That's where I came in. As a scientist and engineer, I spent a few months at the Los Alamos National Laboratories in New Mexico, where the first atomic bombs had been created. Our job was to help the government determine the fastest, most economical, least risky, scientifically most beneficial means of getting to Mars before 2000, possibly in cooperation with the Russians. It was then my hunches were confirmed that Mars 1999 and PhD were the way to go. Little did I realize, though, that I would be going there!

After much deliberation and debate, we prepared four Mars mission options for consideration by the White House and top NASA management. The first was a slow marriage with the Russians that would begin with a cooperative unmanned Mars rover and sample return mission in 1996 and build up to a possible joint human mission there as early as 2010. This was favored by the old guard, who feared a more expansive commitment would not get the necessary public support; they also thought that NASA needed to concentrate on the growth of its space station and transportation system and not go tooling off to Mars so soon.

The second option was to send people on a one-year, fast (hyperbolic) flyby of Mars in 1999 or 2001. This could be done with or without the Russians. The cost was relatively low, but very limited science observations and no propellant processing could take place because, like a speeding train, the crew would simply swing by Mars swiftly in less than one day and spend no time on the planet or its moons.

A third option proposed a U.S.–Soviet joint manned Mars landing and orbiter in 1999 but with no propellant processing setup on Phobos or Deimos. This option optimized the chance of a Mars landing, but left nothing in place to economically justify returns to Mars. It was also the most expensive option.

The fourth option provided for a joint U.S.–Soviet landing, separate interplanetary spacecraft, and a sixty-day staytime near Phobos or Deimos, where the astronauts would set up a water and propellant processing plant. The crew would also teleoperate unmanned rovers on the Martian surface and screen samples for possible contamination. If all systems were go, if the Martian samples contained no biologically harmful materials, and if the PhD propellant plant was working properly, the crew could fill the tanks of a manned Mars sortie vehicle with propellant and land one astronaut and one cosmonaut on Mars toward the end of their stay. The launch window for return to Earth would bring the crews back in November 1999, in time to celebrate the turning of the millennium and complete nuclear disarmament. Meanwhile, tanks on PhD would be accumulating water and fuel for revisits. Three more human visits within the next five years could provide for returning enough propellant to Earth orbit and the

Moon to pay for the investment and allow for the rapid buildup of both lunar and Mars bases. This was Mars 1999, and it was a winner for NASA and the White House.

It was a winner for the Russians, too. It turned out they had also been planning to do this mission. On the very day our study began, the Soviets launched the second of two unmanned spacecraft to Phobos, to sample the surface before sending people there. Their most important goal was to confirm evidence that water was in its soil. That next year, we learned that the water was there: 1989 was the year of discovery that provided the needed proof to convince the skeptics.

The President had little to lose and all to gain from the Mars 1999 proposal. The United States and the Soviets would build separate interplanetary spacecraft and arrange for them to be launched with international crews from Earth orbit in January 1998 on a trajectory past Venus for a December 1998 arrival. Each nation could go it alone if the other backed out, but the presence of the other spacecraft nearby throughout the twenty-two-month round-trip voyage could provide a backup for either ship in trouble, and allow for many crew exchanges during the tedious journey, providing international goodwill along the lines of the 1975 Apollo-Soyuz linkup. Only the Mars landing itself would require joint U.S.–USSR participation.

With prospects for a level NASA budget at 1 percent of the federal budget and sharing the resources with other countries, the President had a winner. He could take credit for broaching this symbol of peaceful cooperation and presenting a pilot project for world government to Gorbachev, who until then had seemed to be stealing the limelight with his disarmament proposals.

The President's hook was applying the model of the U.S. Constitution to world affairs. His statesmanlike initiative was perfect timing for the forthcoming Republican convention, while the Democrats struggled to advance even bolder proposals for world peace and cooperation. It became an international bandwagon.

I recall the President addressing a joint session of Congress. "The greatest adventure of all mankind is upon us," he said. "We must ratify SALT III providing for complete nuclear disarmament by the year 2000. I must caution you, however; this agreement is delicate and requires strict adherence to inspection procedures throughout the coming decade. I am gratified that we are moving in the right direction. To back the spirit of our new accords, and to fulfill our grandest dreams, we must also press on to Mars. Mr. Gorbachev and I have signed a joint resolution that we will do it cooperatively. But if the Soviets back out, we should do it alone. And I assure you, if we don't do it, the Soviets will. As we enter the twenty-first century, the nation that dominates space will dominate the world. We shall not be number two!

"Therefore," he continued, "in your budget appropriation this year, you will be asked to supplement earlier authorizations for certain long-lead-time developments that are essential to the Mars 1999 project. For example, we will need to test in orbit a Mars moon propellant processing plant, which will save us tens of billions of dollars by the year 2005. We will need to test the effects of variable gravity on the human body. We will need to test aerobraking in the Earth's atmosphere. And we will need to rethink the space station in a new role as a base camp, or stepping-stone to Mars. The tradition of American freedom, as embodied in our Constitution and the future of our children, is best represented by action steps like the Mars program that will assure us of a positive future."

4

The Soviet Factor

The scenario of the previous chapter is the riskiest because it occurs shortly after this book's publication and also because it involves controversial issues such as SDI, disarmament, and world government that may or may not affect the decision to go to Mars. Still the scenario symbolizes the kind of political catalyst that will probably be necessary to pull off a Mars mission before 2000. The scenario was written to apply to either President Reagan or his successor.

We can wait a while longer, but not much. If Mr. Reagan chooses not to go for the brass ring, his successor probably will. Meanwhile the letter to the President reproduced in the back of this book remains unanswered, by him anyway, although I did get an unsatisfying response from a NASA official.

One needn't be a cold warrior to surmise that the Soviets have the same goals in mind. As in a tortoise-and-hare race, the United States has gone into a sort of slumber since the crowning achievement of Apollo. Meanwhile the Soviets plod forthrightly along with their bigger space stations, boosters, and shuttles, their endurance records (currently 237 days, the length of a one-way mission to Mars), and their 1988 plans to send a probe to Phobos.

We consistently hear of the Russians stating they want to send people to Mars, but the time frame varies. Soviet Academy of Sciences president Anatoly Alexandrov has proposed as a Soviet goal a mission to Mars with a large cosmonaut crew at the time of the 1998–1999 opportunity. Former Apollo astronaut and U.S. Senator Harrison Schmitt has suggested the Soviets may be on a bolder timetable: 1992, which coincidentally is the seventy-fifth anniversary of the Bolshevik Revolution and the five-hundredth anniversary of Columbus's discovery of the New World.

Other observers are more conservative, but we don't really know how much these statements reflect official Soviet policy. Pilot-cosmonaut Konstantin Feoktistov believes that cooperative programs will need to be enacted step by step, although it would be "entirely feasible to implement in ten to twenty years."* Cosmonaut George Grechko, in a meeting with former astronaut Russell Schweikart and school children from Maine who had proposed the joint trip to Mars, replied, "As for us cosmonauts, we're prepared to sign your proposal here and now. We agree to fly to Mars together. I'm positive that our decision will be backed by our youth, scientists, and engineers, by all people in the Soviet Union for that matter. But before that, we should join efforts to end the cold war. All Star Wars projects should be done away with; otherwise there will be no one left to fly to Mars."

Nikolai Ryzhkov, chairman of the Council of Ministers of the USSR, proposed to the United Nations a stage-by-stage program of "joint practical actions in peaceful exploration of outer space [with the] aim to lay, before the year 2000, solid material, political, legal, and organizational 'Star Peace' foundations."**

The consensus among most current space leaders in both countries is that it is not only desirable to cooperate, it may be necessary. Roald Sagdeev, director of the Soviet Academy of Sciences' Space Research Institute, said a manned Mars mission "would not be realistic without international cooperation, and such an effort may be opened to the world's other space powers."***

---

*From *Leningradaskaya Pravda*, January 11, 1986.

**Press release No. 96, June 13, 1986, available from the USSR Mission to the United Nations, 136 E. 67th Street, New York, NY 10021, and cited in *Whole Earth Review*, Winter 1986.

***Reported in *Aviation Week and Space Technology*, March 26, 1986.

This view was echoed by NASA administrator James Fletcher.* "Human exploration of Mars," he said, "has now moved from science fiction to serious consideration by a presidential commission." He felt the manned Mars mission would be such an enormous undertaking it would need to involve the cooperative effort of the United States and the Soviet Union, as well as other nations.

"Mars is not going to be settled as a national enterprise," said Thomas Paine, chairman of the National Commission on Space.* "Indeed that would be grossly unfair to humankind as a whole. Everyone will want to participate, and it is up to us to provide the leadership. The banding together of all humankind cannot but help to bring our own home planet many benefits [including] a more peaceful, human planet Earth." Soviet journalist Y. A. Golovanov wrote in *Pravda*, "I would like to write a reportage about how the American shuttle docks with the Soviet Salyut . . . and how Soviet and American lads leave the first footprints on the red sands of the Martian deserts."

Senator Spark M. Matsunaga of Hawaii wrote for *Omni* in 1984: "An international [manned] Mars mission . . . would enlist the world's best scientists and engineers in a united enterprise, the most stirring undertaking in human history. . . . While pursuing this new path of exploration and discovery Americans and Soviets may even rediscover their common humanity."

Two of the most ardent supporters of a joint manned Mars mission are Bruce Murray and Carl Sagan, codirectors of the Planetary Society. Murray has said in *Issues in Science and Technology*, Spring 1986, that we need long-term goals from the President now, including "a joint manned mission to explore Mars after the turn of the century."

"Is there not some special obligation," wrote Sagan in the February 2, 1986 issue of *Parade*, "of the two principal spacefaring nations—the two nations that have burdened our planet with 55,000 nuclear weapons—to put things right, to use this technology for good and not for evil, to blaze, on behalf of every human being, the trail to Mars and beyond?"

Representative George E. Brown, Jr., of California is also enthusiastic. He proposed an SCI program, or Space Cooperative Initiative, whose principal focus would be a joint U.S.–Soviet trip to Mars. "It could help demonstrate," he wrote for the *Los Angeles Times*, November 30, 1986, "the un-

*Reported in *Aviation Week and Space Technology* in 1986.

assailable yet overlooked fact that the superpowers have no choice but to coexist on Spaceship Earth."

In his book *The New Race for Space* (Stackpole Books, 1985), James Oberg, author and Soviet space watcher, wrote, "If the Soviets and Americans and other space powers have learned through long practice how to productively cooperate in space, nothing—not even Mars—could be beyond their synergistic reach."

Short of the full-blown U.S.–Soviet human missions, discussions are going on in high places in the United States encouraging a collaboration with the Soviets. Lew Allen, director of the Jet Propulsion Laboratory, endorses a joint unmanned Mars rover and sample return mission in 1996, as reported in the January 26, 1987 issue of *Aviation Week and Space Technology*. The United States would provide the roving vehicle while the USSR would supply a lander with a sample return ascent stage. Sagdeev agrees. An unmanned mission to return samples from Mars "would be a perfect lead-in to the major mission for a manned expedition to Mars," he said.

In 1985, the State Department informally asked the Soviet Union whether it would be interested in a joint Salyut-shuttle mission. The Soviets replied that, because of SDI, the time was not ripe. In early 1987, White House aides and American space officials told *Aviation Week and Space Technology* magazine that if the Soviet Union agreed with such an exercise, "the effort could lead eventually to joint exploration of the Moon and possibly a joint U.S.–Soviet manned flight to Mars."

One Soviet concern repeatedly aired has been the U.S. involvement in SDI. In recent talks the two nations have been able to sidestep the issue and still provide for some cooperation in such projects as a Salyut-shuttle linkup, exchanges of data on forthcoming Mars missions, and the Mars sample return mission. But the issue of SDI still keeps coming up, and Americans are deeply divided. Carl Sagan and Congressman Brown would like to see the potential SDI weaponry converted to a Space Cooperative Initiative (SCI), a view shared by the Washington-based Institute for Security and Cooperation in Outer Space. I too would like to see the space swords-to-plowshares scenario play out.

It appears that a national or international consensus on a Mars mission transcends SDI politics. It stands on its own as a positive and politically bipartisan beginning. Harrison Schmitt, for example, sees a human mission to Mars as an urgent national goal whether or not the Soviets wish to cooperate.* He suggests it be unlinked to SDI. He warns that if we don't go to Mars, the Russians will. They won't wait for us.

Former White House science adviser George Keyworth appears to have become a Mars convert too. "Now is the time for bold, long-range thinking," he was quoted by *Aerospace America* in January 1985; "the public is in the mood for it. One challenge well within our technological grasp is a manned Mars mission." Keyworth felt we might want to sidestep the Moon and go to Mars first. Congressman Brown agrees. "The technical basis for a manned Mars trip," he wrote in his *Los Angeles Times* editorial, "is stronger than the technical basis for President John F. Kennedy's 1961 decision to land a man on the Moon."

In a December 6, 1986, editorial, the *Los Angeles Times* editors stated, "The value of a joint mission to Mars would be so great that it is worth doing the work to make it happen. But if it cannot be done jointly, it is worth doing alone. The United States should be making plans for a trip to Mars. President Reagan should declare this a national goal, and NASA should get to work."

The *New York Times* has also jumped on the joint Mars mission bandwagon. A February 12, 1987, lead editorial noted, "Mr. Reagan has endorsed two bad ideas, the space station and sharing Star Wars technology with the Russians. Going to Mars with them would give NASA a goal worth aiming for."

A new Washington-based lobby group called Search for Common Ground has taken the joint manned Mars initiative as one of its key projects. This group is looking for ways to cooperate with the Soviets regardless of how individuals or governments may feel about SDI and other controversial projects. "Important to our work," a 1986 newsletter stated, "is the ability to anticipate issues that fire public imagination. In recent months, we seem to have found one—specifically, U.S.–Soviet space cooperation leading to a joint manned mission to Mars."

Even the Congressional Office of Technology Assessment, usually conservative in such matters, has joined the Mars bandwagon. In a 1984 report on civilian space stations, the OTA wrote, "In 15 years, NASA's complete infrastructure aspirations and a lunar settlement could be in hand, and

---

*"The Politics of Mars." White paper for the Mars Workshop, NASA Marshall Space Flight Center, May 1984.

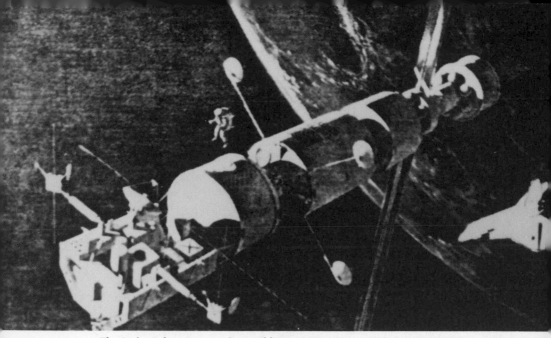

The Soviet *Salyut* space station could grow into a Mars vehicle over the next few years.

perhaps also, plans for seeing a human crew travel to the vicinity of Mars and back could be well advanced."

Of one thing we can be sure. There is intense interest in both the Soviet Union and the United States in sending people to Mars. Regardless of political persuasion, those American leaders that have chosen to look at the possibilities believe we should try to do it jointly. In this country the notion of going to Mars appears to be as unassailable as motherhood and apple pie. It is ready made for an American administration to grab and run with.

Soviet intentions are more difficult to read. Perhaps their more conservative reports suggesting that manned Mars missions would take place much later are a diversion to their real plans for the 1990s. Certainly the relevance of their achievements in Earth orbit cannot be ignored. No other mission except going to Mars could motivate the Russians to keep pushing their endurance record in space, now at close to nine months, or to close their life-support systems as much as they have. A recent five-month experiment showed they are able to recycle 90 percent of their waste products.

Imminent developments that feed into the potential Soviet human Mars infrastructure are the first successful flight of their large booster, bigger than the Apollo/Saturn V; the advent of their reusable shuttle vehicle; longer stays in orbit; a nuclear upper stage; expansion of their Salyut 7/Mir space station; the Phobos 1988 mission; and many, many more potential surprises.

"While the nation is taken up with the aftermath of the *Challenger* tragedy," wrote National Commission on Space member David C. Webb in the March 7, 1986 *Los Angeles Times*, "an event of great significance has occurred that places us ten years or more behind the Soviet Union in space development." Webb was referring to their expanding habitation of space, well ahead of the U.S. space station program.

My own conclusion is that the Soviets may take a crack at Mars 1999 whether the United States chooses to do it or not. Like most major Soviet initiatives, the manned Mars program has not been publicly announced. But it would be silly to dismiss what they are doing as merely exercises in infrastructure. The Russians are surely headed toward Mars, and we can surely do it together.

In his closing statement in the November 1985 summit conference, Secretary Gorbachev said President Reagan had not discussed what Gorbachev identified as the most important issue for the superpowers: how would we go into space, peacefully or belligerently? To some degree this may be an attempt to make the Soviets look good, but it is still an important issue for *both* sides to address.

In an innovative book commemorating our entry into the third century under the American Constitution, New York attorney Joanne Gabrynowicz sees world government emerging from seemingly clashing national interests. Just as the American founding fathers were able to conceive of a supersystem (the Constitution) to bring together widely differing factions, a new global system will bring together nations. The focus, the role model for doing all this, Gabrynowicz writes, is the exploration and development of outer space, and the Mars mission becomes the perfect example.

"An eighteen-month international human mission to Mars and back," she continues, "can capture world attention for a long enough period so that people will begin to believe that global cooperation is, in fact, a viable alternative to global destruction."*

Senator Matsunaga put it this way in his 1984 *Omni* editorial: "An international Mars mission would be the birth of a new, transcendent age. Mankind's highest aspirations in art, science, and religion would be finally realized. The beast of divisiveness would be securely caged as humanity began a journey into the cosmos, unified at last."

---

*"Space Development and the American Experiment: A Context and Model for International Progress." Paper presented at the Princeton-AIAA Conference on Space Manufacturing, May 1987.

A cooperative voyage to Mars is more than a symbol or a handshake or a fleeting moment of brotherhood and sisterhood. In my opinion, it is a positive and focused goal that, more than any other single activity, will bring the planet together. This is the most compelling political, human motivation to launch Mars 1999. And the United States, as the world's social, political, technical and economic leader, has a golden opportunity to lead the world into successful international cooperation as no other nation could.

5

Scenario

Launch from Earth and Swingby of Venus

January through July 1998

• *JANUARY 24 — The U.S. crew boards the space shuttle* Atlantis *for an orbital rendezvous with the U.S. space station and the Mars interplanetary spacecraft, the* Olive Branch. • *JANUARY 27 — After a three-day checkout, both the U.S. and USSR Mars spacecraft escape from Earth on a trajectory toward Venus.* • *FEBRUARY 6 — First U.S.-USSR crew exchange and misunderstandings.* • *APRIL — Resumption of crew exchanges.* • *JULY 4 — With a Venus swingby, the two spacecrafts' trajectories are redirected toward Mars, saving time and fuel.*

T minus fifteen minutes and counting. All six of us strapped in as passengers on middeck of the space shuttle *Atlantis* looked at one another with an unspoken anticipation of the adventure to come. The *Atlantis* would take us on the brief trip to Earth orbit to rendezvous with the space station and the *Olive Branch*.

We had trained together for over a year now, putting the *Olive Branch* through its paces, rehearsing our chores, training on the ground, training in space, moving through life in a group of six acting as one. The friendships and group energy we felt went beyond the stereotypical platoon-at-war fellowship. We had taken seminars on inner awareness, group awareness, risk taking, accountability, honesty, goal setting, and applied human relationship (loving). Even though the crew makeup involved mixed sexes, nationalities, and professions, we were trained to be aligned.

T minus fourteen. Eye contact was warm among us, except for Busby. He seemed tense. Something was bothering him, but I couldn't tell what.

57

Up until that point, I had no reason to suspect any lack of competence on his part. In many ways, I envied him. He had the right stuff and I didn't. I think he had probably wanted to be commander of the *Atlantis* even on this brief forty-eight-minute flight to the space station before assuming command of the *Olive Branch.*

I looked over toward Busby and smiled. In all, Busby had commanded seven shuttle flights over the past nine years. He was the most seasoned of the veteran astronauts. He was the choice of the astronaut office at the NASA Johnson Space Center in Houston to pilot the *Olive Branch.* In fact if the Houston flight crew operations folks had their way, they probably would have wanted to clone Busby five times for projected optimal performance. Busby was a textbook shuttle pilot and experienced test pilot. He had also headed the astronaut office at the Johnson Space Center, a post some critics had said allowed him to appoint himself for the Mars mission.

T minus eight. Marla Lee, deputy commander of the *Olive Branch,* was as conscientious a person as I have ever known. Quietly supportive, she also had the aura of leadership and a warm sense of humor. She was forty-two, single, and attractive, a Ph.D. in astronautical engineering, an M.D., and an artist. We had in common an interest in spirituality—meditation, the New Age, Eastern philosophy—and we shared a curiosity about the latest happenings in the world of UFOs and the quest for contacting life in the universe.

T minus six minutes, thirty seconds. Then there were the Steinmullers, Eldon and Helga. Married for twenty-six years and in their mid-forties, they were a stable German couple. Eldon was half British and had been raised in London. They were sweet and agreeable, though a bit rigid at times, and they were rarely apart. Our group trainings had helped them feel comfortable mixing with the rest of us. As I mentioned earlier, Steinmuller had been the genius of PhD water and propellant processing. If anybody could make that plant work, it would be Eldon. Helga was an M.D. and would also assist Eldon in deploying and operating the plant. Eldon's deal with NASA was, "I'll go if Helga comes." The European community applied enough pressure to NASA to make that happen.

T minus five minutes. Yamoto Osho from Japan was our ship's engineer. Quick, spry, and alert, he was the *Olive Branch's* software analyzer, module replacer, and handyman. He embodied the Japanese success story. Starting out as an auto assembly worker at a Mitsubishi plant outside

Tokyo, he rose to become chief engineer in only twelve years while completing a master's degree in mechanical engineering with highest honors. He joined the astronaut corps shortly thereafter. Japan had made many of the components of the *Olive Branch*. During our year of training, Osho's English had become almost perfect. I always enjoyed his smile and his enthusiasm. He was only thirty-eight years old but very wise for his years and very, very practical.

T minus two. Then there was me. Planetary scientist, astronautical engineer, mission conceiver, pianist, photographer, philosopher, and the oldest man on board. My job was to monitor the unmanned rovers on Mars and to crosscheck other mission functions to make sure they were performing. I was third in line to land on Mars, behind Busby and Lee, and so very little of my time in training had been spent on that part of the mission. Little did I know what would unfold.

So there we were, the *Olive Branch* six, waiting on the launch pad at the Cape. Cross-trained, motivated, intelligent. Negative thoughts of the *Challenger* darted through my mind. What an ignominious way to go, I thought, if we crashed within the first few miles of a half-billion-mile journey almost twice around the Sun and past the two nearest planets to Earth. It would be like having a freeway accident on the way to a ceremony to receive the Nobel Prize. I had trained myself to deflect thoughts like this almost as soon as they came in.

T minus ten, nine, eight, seven, six (I heard a muffled roar of the main engines firing), five, four, three, two, one. A gentle lurch and then some giant lurches as the solid rocket boosters lit fire, and then we were on our way. Slowly we mashed back into our seats. For two minutes we shook as if in a roller coaster, amidst the roars of loud crackling, pressed back onto our chairs. Then a sudden lurch forward, and then a gentle, soft glide as the solids were jettisoned. Still mashed in our seats, we accelerated toward the staggering speed of Mach 24.

T plus eight minutes. Now the shuttle's external fuel tank was jettisoned, and we once again lurched forward, this time into weightlessness. We were in space. This particular external tank, I reflected, had its fate sealed in the Indian Ocean, where it would soon splash down. The ET we would rendezvous with in orbit, which we called *Ironhorse*, was probably the most durable and economical device ever created by man. Not only had it been a booster from Earth to orbit, it had carried 130 extra tons of fuel aloft, received another 400 tons of fuel from other ETs, would launch us from NASA's multi-modular space station transportation depot in Earth orbit onto an escape trajectory toward Venus and Mars, would serve as a fuel tank to stockpile liquid oxygen and hydrogen processed from water on Phobos, would in two years return a 1,200-ton payload of water from Phobos back to Earth orbit, and would eventually use some of that water, once processed into more fuel, for another trip to Mars and then back again! What an investment, I thought.

This was only my second trip into space. The first had been our training mission, which kept us aloft for thirty days about seven months ago while we put the *Olive Branch* through its paces, deploying and retracting it to and from the interplanetary (spun-up) cruise mode.* We had simulated Mars landings, unmanned rover teleoperations, PhD processing, and aerobraking at Mars and Earth.

As we floated through the blackness of space, I reflected on how I had originally been selected for this mission—a combination of fate, chutzpah, a little luck, a lot of motivation, and salesmanship. Many thought I would be too old and might keel over with a heart attack. But my selling points stuck: cosmic ray exposure, I argued, was best taken by older astronauts, because the risk of eventually contracting cancer was lowest. I also sold them on the concept of my unique experience with Mars and with the astronaut program.

The six of us docked with the space station at T plus forty-eight minutes. A launch crew on board the *Olive Branch* complex was now making last-minute checks before yielding the ship to the six of us. The mission module, ET, and other spacecraft bound for Mars were co-orbiting 500 meters away from the space station. This crew would soon arrive at the space station on their Orbital Maneuvering Vehicle ferry, a device we would be using to fly to and from Phobos, Deimos, and, along the way, the Soviet craft *Mars 1*. The only thing left to do before blasting off from low Earth orbit toward Venus and then Mars was a two-day final crew checkout of the *Olive Branch*. It was synchronized with a similar process on board *Mars 1*, which was co-orbiting with the Soviet Mir 3 space station in a much different, higher inclination orbit.

But last-minute problems with *Mars 1* had delayed our joint departure one day. On January 27, 1998, my fifty-eighth birthday, the *Olive Branch*

---

*By spinning up the modules at the ends of a tether, we can create an artificial gravity so as to eliminate the physiological hazards of weightlessness.

The NASA space station serves as the base camp for assembling the Mars 1999 vehicles. *Photo courtesy Rockwell International.*

and *Mars 1* escaped Earth orbit and rendezvoused near the Moon, the first step on a historic journey: five and one-half months to Venus, another five months to the moons of Mars, and one year from today, the landing on Mars.

Why go to Venus first? Because, believe it or not, celestial mechanics tells us we can get to Mars and back more quickly and cheaply that way. The gravity of Venus would accelerate us out to Mars rapidly. We would be arriving on a date when it would be convenient to stay for one or two months and then return to Earth. The more direct route to and from Mars requires an extra year of waiting while the Earth gets lined up properly for the return. Every two years, we have a chance to launch a crew to Mars on each kind of trajectory, although the Venus swingbys are not always feasible. This opportunity was. Those are the facts of science, known years in advance.

T plus ten days. Earth and Mars receded into small discs and the coldness of interplanetary space was upon us. The *Olive Branch* and *Mars 1* had been working perfectly so far, and we were ready to host the first visit of four crew members from the Soviet craft. We looked forward to the event eagerly, knowing these exchanges would break a lot of monotony and promote international harmony.

Mars 1 bore little resemblance to the *Olive Branch*. The Russian craft was not spun up so as to create artificial gravity and was very much simpler in design. It consisted of two Mir-sized modules end to end and a large propulsion module. (*Mir* is the name of the Soviet space station of the late 1980s and the 1990s). The *Mars 1* crew would travel to Mars in complete weightlessness. Unlike their American counterparts, the Soviet designers had been convinced that the physiological effects would not incapacitate the crew.

The *Olive Branch*, on the other hand, was a strange-looking assemblage of co-flying spacecraft. The mission module included a heat shield for aerobraking at Mars and Earth. There was a central hub module and docking adapter. Then came two fifty-meter tethers extending in opposite directions inside an inflated passage tunnel that joined two slowly spinning modules in which we lived and worked. The spacecraft spun at two revolutions per minute, giving us the experience of one-third of an Earth gravity in the two modules. Orbital tests had shown that one-third of a g was adequate to prevent bone decalcification. Coincidentally, this was also the gravity on Mars. We would commute from our hab module at one end, through the weightless hub module, where we socialized, then to the counterbalancing laboratory module, where we did most of our work. Quite a commute through variable gravity!

The *Mars 1* crew, on the other hand, would experience weightlessness throughout the mission. Instead of aerobraking at Mars and the Earth and pulling four g's both places, such as a retracted *Olive Branch* would experience, *Mars 1* would perform gentle propulsive maneuvers to insert the crew into Mars orbit and later into Earth orbit. The Soviet propulsion modules were huge, having been launched from Earth by their giant booster.

I already mentioned the results of our first crew exchange. Two Soviet cosmonauts, one Bulgarian cosmonaut, and one Indian cosmonaut boarded the *Olive Branch* on February 6, 1989. Everybody joined in a circle inside the weightless hub module. Sevastyanov opened a bottle of vodka, poured it into ten covered squeeze bottles, passed them around, and proposed a toast.

"Comrades," he said in broken English carefully rehearsed and with an enthusiastic flair, "I celebrate the joy of our brotherhood on this momentous occasion. May we all have a safe and friendly journey."

"I'm sorry," said Busby, handing his squeeze bottle back to Sevastyanov. "I can't allow my crew to drink. It's against regulations."

After a long, awkward silence, Lee said, in Russian, something about consummating our appreciation of friendship even though the commander would not permit the drinking of vodka, and she hoped they would understand.

But the damage had been done. The morale of both crews drained as though suddenly drawn into the vacuum surrounding the spacecraft. The voyage to Venus became tough. All crew exchanges were cancelled. And when the reaction control system of *Mars 1* became stuck and the spacecraft began tumbling, offers of help from the *Olive Branch* were coldly refused (they solved it, thank God). Then other things started to go wrong. The *Ironhorse* and its precious cargo heading to Mars and Phobos alongside us had a tumbling problem. The astronomy experiments Lee and I tried to perform all failed. The refrigeration unit was out. Even the optimistic Osho could not find the solution to the problem. It was as if our collective human failure precipitated machine failure.

Being unable to perform experiments, we were faced with hours of idle time. During those months, I spent a lot of my time alone in the hub module doing my personal inner awareness exercises (or sadhana, as it's known in India). In the depths of space, I found myself asking for God's help to restore the morale of both crews. It seemed incredible that we human beings, carefully selected for our intelligence, achievement, and psychological fitness, were having trouble getting along. Oh, the vulnerabilities of the ego!

All those handy communication skills we had learned seemed to go out the window. The Steinmullers retired to their corners; Busby convened tense and unproductive staff meetings, which Lee rarely attended; and the smiling but vulnerable Osho was being courted and manipulated by Busby, who at one staff meeting designated Osho deputy commander. I had trouble communicating even with my good friend Lee, who spent most of her time in her cabin.

Crew exchanges did start up again after two months, thanks to diplomatic guidance from Earth. Skilled psychologists and negotiators helped us find areas of common interest, although the widening communication gap with Earth made it awkward. As the months passed, we were already a few light minutes' communication time away, making these interplanetary counseling sessions not all that useful. We began to realize we had to mend our differences ourselves.

I can't recall any one event that began the healing of our group illnesses, but there was one very funny videotape the Soviets showed us about Venus. During the 1960s and 1970s the Soviets had landed on Venus a number of unmanned spacecraft called Venera, some of which managed to survive the oppressive heat (400 degrees Centigrade, hot enough to melt lead) and atmospheric pressure (100 Earth atmospheres). Part of the videotape was an animation of hypothetical Venusians greeting the spacecraft and extolling the virtues of Soviet society as if they already knew it. It was cheap propaganda. It was also hilarious—unintentionally. Watching it simultaneously, crews from the *Olive Branch* and *Mars 1* broke into unstoppable laughter.

We then speculated about encounters with the Venusians during the flyby, dusting the ubiquitous cloud cover with hammer-and-sickle emblems. "This is your planet, Comrade," I would say amidst gales of laughter. "You are to get off here and we'll go on to Mars! But really I'd like us to go together." The Russians got even by playing back portions of cold war speeches by American presidents, including "We are number one." We laughed.

The tension had lifted totally by the time we flew by Venus on July 4, 1998, even though our laboratory module was overheated. The planetary encounter was awesome. Swinging only 2,000 kilometers above the clouds (closer than Phobos is to Mars), this huge featureless yellowish-white crescent filled the windows. We snapped pictures and reflected on the awesomeness of the first human encounter with another planet.

What a gift from nature, I felt, that Venus was thrown in as a bonus on this mission. It was also a great help. Thanks to its gravity assist, we would go to Mars and return within two years. Without Venus, it would have taken three.

6

Blueprint for Mars 1999

Some of the pieces of the Mars 1999 puzzle are now probably clearer. But most readers are not celestial mechanics or astronautical engineers, so more explanations are in order. Many concepts seem to violate common sense.

First, what is the easiest way to get to another planet? On Earth the best way to get from A to B is a straight line (or curved line, really, around the circumference). In space it is very different. Generally speaking, travel in space involves a rocket in a given orbit briefly imparting thrust to a payload of cargo or people in such a way that the payload coasts in a new orbit that intersects the orbit of the target object—when that object is there.

Newton's laws of motion allow us to calculate that the least amount of rocket fuel will be needed when our spacecraft orbit (say, around the Sun, joining the orbits of Earth and Mars) starts with its closest point to the Sun (perihelion) at Earth departure, and its farthest point from the Sun (aphelion) at arrival upon Mars. The spacecraft makes one-half of its orbit around the Sun and the one-way mission takes about nine months. This is called a

Hohmann transfer. Opportunities to go to Mars, when the planet happens to be there at the spacecraft's aphelion point, occur every two years (or one Martian year). That's how the planets line up.

The only problem is, when you get to Mars, you have to wait another eighteen months or so for the planets to line up properly to do a second Hohmann transfer back to Earth. These missions are optimized for the minimum use of fuel, but a three-year total mission time may be too long for the comfort and safety of the crew, especially on the first mission.

The way around the problem is the Venus swingby, which involves a Hohmann transfer first to a precise point in space near Venus such that the trajectory of the spacecraft is redirected toward Mars. This crack-the-whip maneuver obeys the well-known conservation laws of physics. The Venus

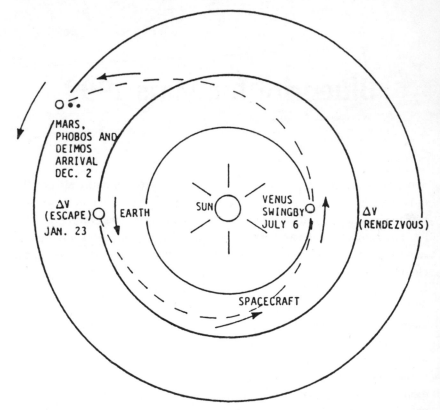

Minimum-energy transfer from Earth to Mars, Phobos, and Deimos with Venus swingby: 1998.

swingby chance recurs every two to three years either outbound toward Mars—which is what Mars 1999 is all about—or inbound from Mars. These opportunities are conducive to Mars layovers of up to about sixty days and total mission times of somewhat less than two years. Going by way of Venus saves time.

The main problem with the Venus swingby mission is that a large amount of energy is sometimes then needed to brake the spacecraft at Mars (or at Earth), because the spacecraft, after being redirected by Venus, usually comes in at a very high velocity. The way out of that problem is to bring along a heat shield and use the upper atmosphere of the destination planet to provide enough friction to put on the brakes rather than using up valuable propellant. So the major two alternatives are a three-year direct-to-Mars round trip and a two-year Venus swingby round trip.

A third alternative mission to Mars takes only a year and involves a high-speed flyby of the planet and prompt return to Earth. You can't stop over at Mars or in orbit, however, because the planet's gravity is used to swing you around back toward Earth (a kind of celestial crack-the-whip). This mission could be used to drop off or pick up payloads, but it isn't at all useful for a self-contained mission involving Mars surface science and PhD process-ing—which is what this book proposes.

In all cases, missions to Mars involve four major events, measured by a parameter called delta-v, or velocity increment. The first event is a propul-sive blast in Earth orbit that places the spacecraft on an escape trajectory toward Mars, perhaps by way of Venus. This is called Trans-Mars Injection (TMI).

The second maneuver (even if Venus is used) calls for putting on the brakes at Mars and capturing the spacecraft into a Mars orbit (for example, at Phobos or Deimos). This can be done either by aerobraking or by applying a propulsive thrust. This is called Mars Orbit Injection (MOI).

The third maneuver is blasting off from Mars orbit on an earthward trajectory, called Trans-Earth Injection (TEI). The fourth consists of putting on the brakes, either propulsively, by aerobraking, or possibly with a gravity assist from the Moon. The spacecraft ends up in a rendezvous with the space station. This is called Earth Orbit Insertion (EOI).

Engineering data give us a good idea of the relative merits and draw-backs of each mission. The two-year Mars 1999 mission with aerobraking at Mars and at Earth looks best, for a number of technical reasons beyond the scope of this book.

A similar mission opportunity occurs in 2001, this time with a Venus inbound swingby; but we would need more fuel for this mission. The season on Mars will be much more likely then to spawn huge dust storms that could obscure views of the planet and make landings hazardous if not impossible. Solar flares are most dangerous during this time. And the return to Earth would no longer be in November 1999, the dramatic eve of the third millennium. Also, as we shall see, Mars 1999 can trigger a sequence of follow-on missions that would make rapid and economical use of PhD water and propellants.

Mars 1999 will use existing hardware and technology wherever possible. It is a conservative scenario, still challenging, but do-able. If new technologies come along, so much the better. More on that later.

The first step will be to assemble in Earth orbit—probably near the space station—the components of the spacecraft that will go to Mars.

Engineering studies show that the payload leaving Earth orbit toward Mars would weigh 200 metric tons or more. The fuel needed to boost that payload from Earth orbit on an escape trajectory toward Venus would weigh about 400 metric tons. So we need to get 600 metric tons, or 1,300,000

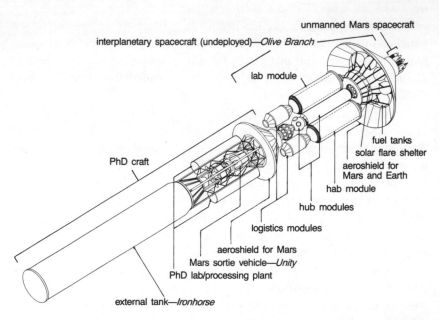

Artist's concept of the Mars 1999 vehicles before Earth escape. *Drawing courtesy SCI-ARC.*

pounds of material, from the surface of Earth up to the space station—no small task.

One simple way I am proposing is to slightly modify the space shuttle system. If we simply take the shuttle off and refit its engines onto its large external fuel tank (containing liquid oxygen and liquid hydrogen), we now have a lift vehicle able to launch about 100 metric tons to the space station. Then a total of six launches of payload and propellants, plus one normal shuttle launch for the crew, would be all that is needed to assemble components and load up propellants.

I am suggesting also that one external tank (ET) itself be used as a booster from Earth orbit to PhD, as a PhD storage tank, and a PhD return vehicle. Propellants from other ETs would have to be transferred into the Mars-bound one, which would be specially equipped with insulation around the tanks so little of the very cold liquid oxygen and liquid hydrogen would boil away. About one-half of the 700-metric-ton propellant capacity of that ET would be needed.

I envision the Mars vehicles made up of three sections that would separate after Earth escape. The front section consists of a number of un-manned spacecraft, including (1) several Mars surface rovers, some equipped with sample return ascent rockets, with the landers targeted for direct entry through the Martian atmosphere to preselected sites on Mars; (2) Mars orbiters that would provide communication links between the surface sites and command centers on PhD, Earth, and aboard the inter-planetary spacecraft that will be located in Mars orbit near Phobos or Deimos; and (3) one or more orbiters that would provide continuing global pictures and other data of the Martian surface.

The second section is the interplanetary spacecraft, or what the engi-neers sometimes call the mission modules. This is what will be making the round trip from Earth to PhD and back. The front of the spacecraft is a heat shield that will provide thermal protection in aerobraking both at Mars and upon return to Earth, and solar-flare protection for the astronauts. During the rocket blasts at Earth escape, crew modules are nested behind the aeroshield. Later two of them will deploy and begin to spin up at opposite ends of a tether to create artificial gravity for crew members during the interplanetary voyage. A third module remains at the hub of the spinning conglomerate.

The third section includes the refueled ET, the Mars lander (what the engineers call the Mars Excursion Module, or MEM), and the PhD labora-

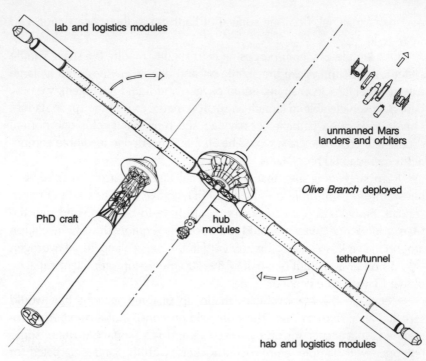

lab and logistics modules

unmanned Mars
landers and orbiters

Olive Branch deployed

PhD craft

hub
modules

tether/tunnel

hab and logistics modules

Artist's concept of the interplanetary spacecraft after Earth escape. In the spun-up mode it is separated from the Mars-bound unmanned spacecraft and the PhD craft, both nearby. *Drawing courtesy SCI-ARC.*

tory and processing plant. The modules are located behind an aeroshield that will be used at Mars only and will take a one-way trip either to PhD or Mars. All will land on PhD, and the MEM will be fueled from PhD oxygen (and probably hydrogen) for the attempted sortie to the Martian surface toward the end of the sixty-day staytime.

After Earth escape, the three sections separate. The mission modules unfold and spin up with small rocket engines to two revolutions per minute, providing gravity for the crew by means of centrifugal force outward. As we shall see, we will probably need artificial gravity for a mission this long, unless something better comes along. A tether in tension and an inflatable tunnel about fifty meters long connect each crew module with hub modules, which serve as a safe haven, solar flare shelter, social space, and free-fall recreation and quiet space.

The length of the tunnel-tether will depend on what gravity the crew would find healthy. A sixty-meter radius of spin corresponds to a Mars gravity or one-third of an Earth gravity. We do not yet know what gravity is ideal; we won't get any answers until we begin to experiment with tethered spacecraft in Earth orbit during the 1990s. The answer will probably lie between one-tenth and one Earth gravity (anything less than one-tenth gravity turns walking into floating).

The inner hub module at near zero gravity is surrounded and shielded from solar flares by the heat shield and by fuel tanks and rocket engines that will be used to rendezvous the spacecraft with PhD and for the return trip to Earth. Additional shielding will be tanks containing water and other materials. A second hub module farther out from the heat shield will be the first space accessible to visitors from a Soviet interplanetary spacecraft that might be flying alongside.

In all, the spacecraft will have three space station–sized modules, 4.5 meters in diameter and 12 meters long. One of the tethered modules is for sleeping, another for working, and the third, in the center at zero gravity, for eating and recreation. The other modules are much smaller and serve as logistics suppliers, docking adapters, the solar flare shelter, and cupolas for viewing.

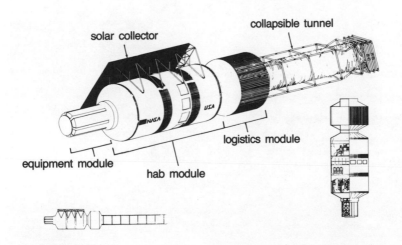

The deployed habitability and logistics modules spin at the end of a tether joining them with hub modules by an inflatable tunnel and truss structure. *Drawing courtesy SCI-ARC.*

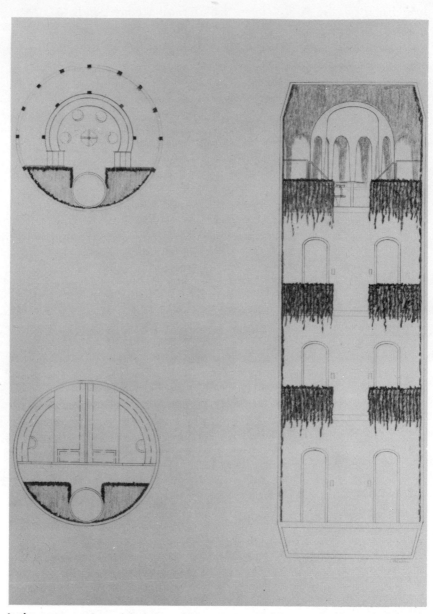

In these cross sections of the hab module, we see two crew cabins per level on three levels and a control center and communal space at the top level (*right*). Floor plans are at left. *Drawing courtesy SCI-ARC.*

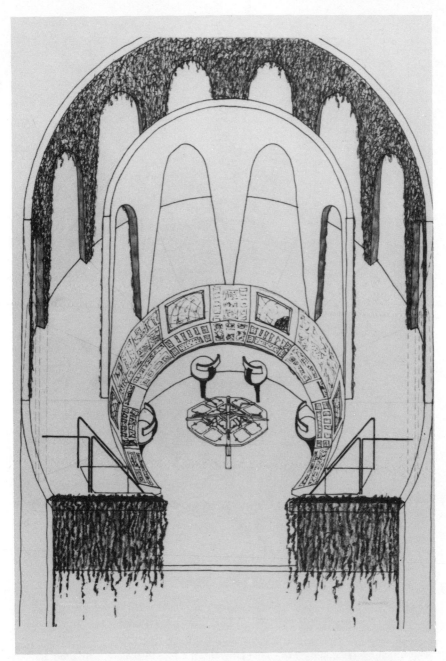

View of the control center in the hab module. *Drawing courtesy SCI-ARC.*

Every time the space shuttle is launched, it is attached to a large external tank (ET) and two solid rocket boosters. In the Mars 1999 program, the ET will achieve Earth orbit and be used to send cargoes to and from PhD.

The external tank in the other spacecraft assemblage that will fly along-side the crew will land on PhD and will stockpile up to 700 metric tons of propellants (liquid oxygen and liquid hydrogen processed from water on PhD) for the next visit to Mars, possibly in late 1999. The external tank will serve as booster to return large payloads of water and other materials to the vicinity of Earth. It could also be used to transport water and propellants to the surface of Mars during a subsequent visit to Mars.

In 1985, workshops sponsored by the NASA Marshall Space Flight Center and Los Alamos National Laboratories gave engineering details of a mission similar to Mars 1999, except that going to PhD and processing for water and propellants were not included. Interestingly, the total weight of payload sent toward Mars is similar in the two cases. We can save forty metric tons of propellant in a mission that would land a crew on Mars if the fuel were to come from PhD rather than be launched from Earth. But approximately an equivalent weight would be invested in propellant to rendezvous with PhD and in the PhD water and fuel processing plant itself.

The benefits of the PhD base are very great. For one thing, we will be processing fuel for subsequent visits that could begin to move large pay-loads of water and other materials (more than 1,000 metric tons) to the Martian surface, the lunar surface, and to Earth orbit by 2001. Second, we can deploy a large network of unmanned Mars exploration vehicles that are controlled from PhD by astronauts in real time with rapid return of scientific data. Third, sample returns from some of these sites to a PhD analysis laboratory will allow astronauts to screen them for possible biological contamination before authorizing the go-ahead for a manned Mars landing.

The only apparent disadvantage to this scenario is the risk that propellant processing at PhD will not work and so the manned sortie to Mars would have to be scrubbed. The trip to Mars is not critical to the survival of the crew. Yet economics mandate a high priority for successfully producing propellants on PhD, so the chance of success is high. Other potential surprises, such as uncertain contamination or failures during MEM check-out, are also possible. If the Mars landing does not succeed on the first mission, there will be a second opportunity later in 1999.

Perhaps the most important element to the success of a human mission to Mars is crew comfort and happiness. Throughout NASA history, the design of manned spacecraft has been dominated by engineers. Their perspective has been clearly utilitarian, and as a result quarters have been battleship gray, high-tech dominated, antiseptic, noisy, quasimilitary bar-

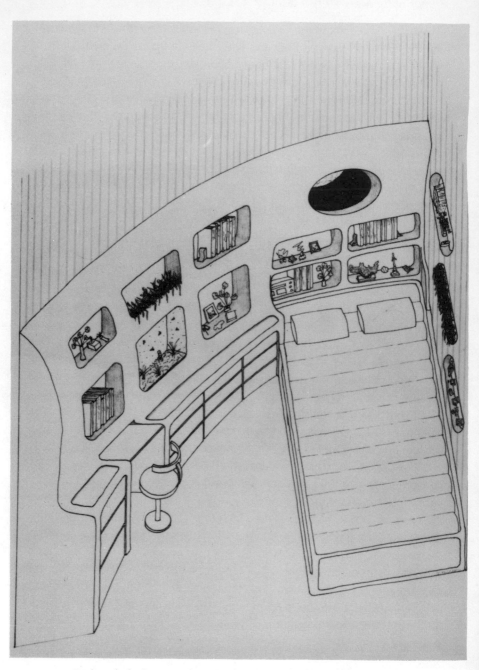

Design of a bedroom on the *Olive Branch*. *Drawing courtesy SCI-ARC.*

Bedrooms can be designed to fit the personality of the crew member. *Drawing courtesy SCI-ARC.*

Dining room, library, and entertainment space. *Drawing courtesy SCI-ARC.*

racks. Aesthetics and spaciousness have been ignored. By and large this approach has worked for the relatively short missions that have dominated the U.S. space program so far. Even the three-month stints planned on board the space station in the mid 1990s will not push human endurance limits in space nearly as much as the Mars mission would. Not only will the space station missions be shorter than the Mars mission, but an unhappy astronaut could always be returned to Earth within a few days.

If there is to be any negative pull to the Mars mission, it would be the engineering community's insufficient awareness of human factors, especially pleasant and safe crew surroundings. Isolation experience in Antarctica, Soviet space experience, and our eighty-four-day Skylab mission all provide examples of the importance of high morale and pleasing aesthetics. If it costs $1 million more to furnish the interior of the mission modules to meet human needs over many months, why not spend it? It would be less than .01 percent of the mission investment. When we think of the sheer

Sick bay. *Drawing courtesy SCI-ARC.*

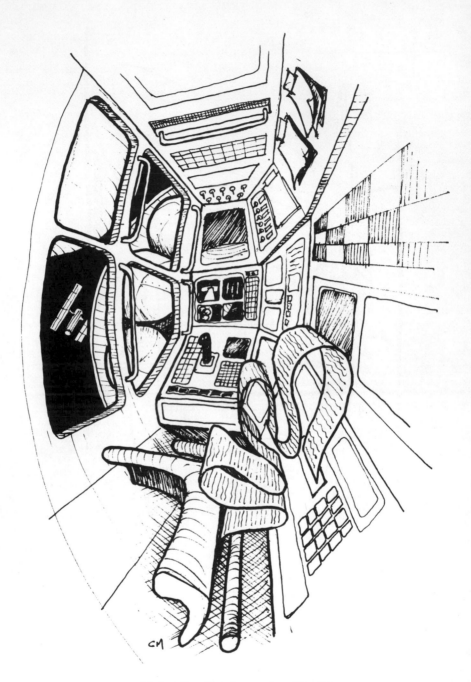

Work station. Drawing courtesy SCI-ARC.

The crew quarters aboard the interplanetary spacecraft can be quite pleasant. *Photo courtesy NASA.*

length of the flight to Mars, going in relative style will be the only plausible way. By comparison, the seven-day shuttle flights are like camping trips and the space station is like a professional scientist's experimental run (perhaps like an astronomer going to an observatory). The Mars mission must go beyond that to become a habitat, a home in space. It shouldn't be difficult to make the spacecraft comfortable.

What can we do to make the mission modules pleasant places to be? First, we need to consider the physiological problems of a mission this long. Most Mars mission scenarios assume a weightless spacecraft floating to and from Mars. I suggest such an approach could be dangerous. We know from the Soviet missions that lasted several months that readapting to Earth gravity after being in conditions of no gravity has created serious difficulty; the cosmonauts have lost bone calcium, even if they exercise in space frequently, and heart problems have also developed.

What if, upon aerobraking at Mars after a one-year weightless mission, the crew members pull four g's and suddenly become basket cases? It would probably mean they couldn't deploy any of the equipment they brought or be able to walk outside the mission modules in space either on PhD or on Mars.

Fortunately, the remedy for this problem is simple: place two of the mission modules at the end of a tether and spin them up. Engineering studies show that the weight of such a tether and spun-up rocket system would be very low (probably less than one ton), and the crew could be provided with safe artificial gravity. The tether would need to be long enough for the crew to experience enough gravity without spinning too fast. A spin of more than two revolutions per minute can disorient a person, just like an amusement park ride, because the inner ear picks up the rotary motion.

It is interesting and curious that no U.S. or Soviet space program incorporates any test of what gravity level is best. This could be done easily with experiments in Earth orbit. We will need to launch the mission module (or a prototype of it) in Earth orbit well before the 1998 launch date to determine the best gravity levels for the flight to Mars.

The NASA space station, planned for 1994 occupancy by a crew, won't help. It is planned to be weightless. But we will need to do some basic research by spinning up a spacecraft on a tether, perhaps in an orbit near the space station. This is the kind of experiment some of my colleagues at Science Applications International Corporation (Buzz Aldrin and Bill Haynes) have proposed as a joint effort with the Soviets. These experiments shouldn't be expensive, and they cannot be ignored if we want to pull off Mars 1999 or any other Mars voyage with success.

Other potential dangers confronting the Mars mission crew are cosmic rays and solar flares. For both, we are concerned about natural high-energy particles and waves. They typically bounce off Earth's magnetic field or are absorbed by Earth's atmosphere, making life safe here and in low orbits. But in interplanetary space, on the Moon, and on Mars, we are exposed to particles and waves that easily penetrate spacecraft shells and over time can induce cancer by damaging human tissue and genetic material. The effects are similar to those created at Chernobyl or what is possible in a nuclear war.

The space shuttle, space station, and Salyut are human-occupied spacecraft that are inside the magnetic belts of Earth and therefore safe, to a large degree. Not so for ventures to Mars, the Moon, and geosynchronous (twenty-four hour) orbits. Studies have shown that the maximum lifetime astronaut radiation dosage should be in the range of 100 to 400 REM, depending on the astronaut's age at first exposure. (REM is a unit of measurement of radiation dosage.) The older the astronaut, the higher will be the allowable dosage; hence the fifty-nine-year-old flying to Mars in our scenario. A younger astronaut exposed to 200 REM has a 3 percent chance of contracting cancer later in life, whereas it is considerably less for an older astronaut.

Mars missions will involve cosmic radiation exposures of twenty to fifty REM per year. This means that, for conservative planning, shorter missions and older astronauts would minimize the risk. Eventually, Mars, lunar, and PhD bases will be able to provide crews with complete cosmic ray protection by covering over their habitats with two meters or more of dust. In Mars 1999 cosmic rays will present a marginal risk, but they are by no means enough to stop the mission.

Whereas cosmic rays are steady, solar flares are sporadic and can be very dangerous on occasion. Fortunately, we will not need as much shielding, even for the worst flares. Advance notice for flares is usually days or hours from the time the sunspots, with which they are associated, first appear. Flares are by far most common near the time of maximum solar activity during its eleven-year cycle. Truly dangerous flares that could kill

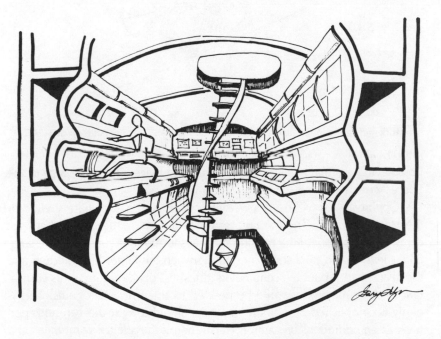

Laboratory module design. *Drawing courtesy SCI-ARC.*

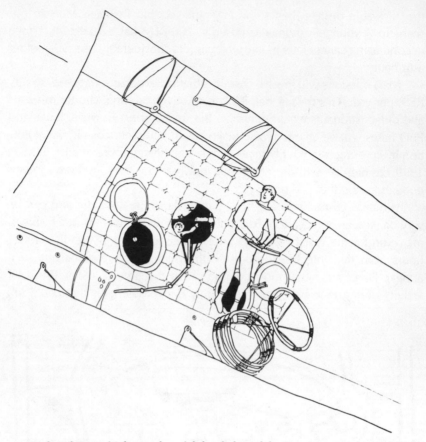

Work and games in the nearly weightless hub module. *Drawing courtesy SCI-ARC.*

unprotected astronauts have happened at least twice since the dawn of the Space Age. These events lasted for a few hours.

Obviously, we will need a warning system and a solar flare shelter for the Mars astronauts. The warning system will require observatories at least on Earth and on board the mission modules to monitor sunspots. The solar flare shelter could be a small module between the larger hub module and the heat shield. About ten centimeters' thickness of aluminum, or its equivalent in other materials, would be needed to shield the astronauts. This is easily accomplished by pointing the heat shield toward the sun and, perhaps as an additional precaution, surrounding the shelter with water and fuel tanks.

Lounge area and community space. *Drawing courtesy SCI-ARC.*

The inner hub module could also serve as a quiet space for crew members to relax and get away from it all. I envision it as spherical in shape with simple white walls and without gravity. It would be like going back into the womb, or like being in a neutral buoyancy tank with sensory deprivation. New York architect Michael Kalil has designed areas like this for the space station. "Zero gravity is like a jewel," said Kalil, "like the sacred space from which we have all come." Kalil sees this retreat ground as psychologically similar to the small chapels adjoining houses in the Italian Renaissance or Japanese tea gardens.

A typical day on board the interplanetary spacecraft might look like this: Awaken in the hab module, wash up, dress, and climb the spiral staircase up through the inflated tube through lower and lower gravity into the outer hub module. Have breakfast with the crew, exercise, and then walk down another stairway to the other module to spend a day at work in the laboratory module. Spend a stretch of time alone meditating in the inner hub module. At the end of the working day, enjoy dinner with the crew, listen to music,

CONCENTRIC CIRCLES OF HUMAN LIFE IN SPACE

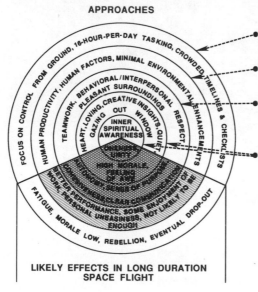

read in the library, and perhaps have an exchange with the crew of the Soviet spacecraft.

The most important considerations in this mission involve making sure the crew has enough comforts to feel at home, enough time for work to avoid stress, and enough opportunities for both private retreats and socializing in the company of other crew members. Ideally, social activities will include the Soviet crew members who would be nearby.

And we need to begin the designing soon. We will need to measure physiological effects of tethered spinning spacecraft on astronauts. We will need to design the space station to accommodate the Mars missions as a top-priority, early function. We will need to test aerobraking at the top of Earth's atmosphere. We will need to rethink and redesign shuttle launches and the external tank. We will need to test in Earth orbit the PhD-destined methodologies for low-gravity water extraction and fuel production. And we will need to change the current NASA and Administration mindset of dwelling on problems rather than moving into bold solutions. The various engineering items are quite manageable with *today's* technology; changing the mindset will be the most challenging.

Mars 1999 can be achieved for $20 billion in 1987 dollars, if it is well conceived and managed. This price is competitive with other human Mars missions that have been proposed, but this one has much higher return on investment, both in science and in fuel production. Most of the funding would go into the research, development, and testing of the various elements composing the interplanetary spacecraft and PhD craft. This could be done at a very reasonable pace during the 1990s within NASA's current budget of $10 billion per year—1 percent of the total federal budget.

But we need to get on with it, and soon.

7

Scenario

Rendezvous with Phobos

December 2, 1998

• *OCTOBER 31, 1998—Command of the* Olive Branch *changes hands.* • *DECEMBER 2— The U.S. and Soviet spacecraft arrive near Mars and rendezvous with Phobos. Remote-controlled unmanned craft begin the exploration of Mars.* • *DECEMBER 5—The fuel-processing plant on Phobos begins its chores.*

In July, our dramatic encounter with Venus had bent our trajectory back toward Earth's orbit and beyond . . . to Mars. During the first half of the year we had headed in toward the Sun. At Venus the total available solar energy was twice that we get at Earth. During the last half of the year, we moved rapidly away from the Sun to Mars, where the available solar energy dropped to one-fifth what it had been at Venus.

This made a big difference in our power budget, which relied on solar collectors affixed to the spinning modules of the *Olive Branch*. That was the bad news about going to Mars: energy conservation.

The good news was getting away from the oppressive heat we experienced near Venus. I already mentioned that many of our experiments were wiped out and that the refrigeration unit in the lab module didn't work. The solar collectors on the sunward side of the lab module, along with metal louvers that radiated away waste heat, were simply not adequate to handle the problem. The heat became so intolerable we closed the lab for two months. It felt like vacation. While residents of Earth enjoyed their summer in warm places, we had ours in a warmer place: near Venus.

By an ingenious series of deductions and brainstorming sessions, our intrepid gadgeteer, Osho, had managed to find a solution to the thermal problem. As in the U.S. Skylab 1 mission, he did an EVA and rigged a parasol outside the sweltering lab module, enough so he could get inside and make the wiring changes he had figured out. It worked, and we were back in business in September, just when earthlings were going back to school and to work.

Even though some of our experiments didn't work, I was grateful for the pause. We had been working very hard. The mission planning engineers had given us more work than we could handle, because they believed we would be bored during the voyage. During the two months we couldn't commute to the lab module, I was glad to have more time for meditation. I also found great enjoyment in playing chess with Sevastyanov, either aboard *Mars 1* or in the outer hub module of the *Olive Branch*. Sevastyanov used to be chess minister of the USSR and I was a beginner, so he usually beat me, but we had fun.

On the *Olive Branch*, we also started to create some experiments of our own along the variable-gravity tubes joining the spinning modules with the hub modules. With computer printouts, tissues, and rubber bands we produced what we called fluff balls; they became a source of joy and communication only the *Olive Branch* and *Mars 1* crews could appreciate. Imagine lobbing balls through the fifty-meter-long inflated tubes where gravity ranged from 0.01 to 0.3 g, with the angular motion as a factor. Imagine taking a magic marker and drawing targets along the tubes, a new spot almost every day. We played space darts, handball, basketball, volleyball, and combinations of them.

When the lab module was fixed, Earth command center began beckoning us back to work there. It wasn't too long before the daily staff meetings became tense again. Busby was on one wavelength, the rest of us were on another. He saw things in a linear fashion, and we were more sporadic in our approach to tasks. The more he attempted to impose his will on us, the more frustrated he became.

About one month away from Mars, on Halloween night, Lee called a meeting with all of us except Busby, in the hub module library. By this point it was clear Busby was isolating himself, grasping at straws to justify his orders, and acting strangely. He called the rest of us disobedient.

"I think the four of you are in pretty good mental health," Lee said. "And I feel I am too. In a month we are going to be very busy people, and I trust we'll all be cooperating to get our jobs done as best as we can. Does anybody have a problem with that?"

We nodded in agreement.

"With that in mind," she continued, "and with the difficulties each of us is experiencing with John, in my position as vice commander, I am taking command of this ship effective immediately. Does anyone object?"

There were no objections.

She continued, "I'd like this transition to be smooth. We have no good protocol to handle this situation. I'm afraid Commander Busby may not understand what's happening or do something rash. We need to be gentle with him and maybe even continue to acknowledge that he is commander, especially if he responds violently. But we will also need to impress on him the importance of getting our tasks done at Mars, and I may be needing to hold frequent meetings with you to get a consensus when conflicts arise and some hard decisions need to be made."

"Can we stay out of this?" Eldon Steinmuller asked. "We have orders to set up that plant and get it working. We have no time for human foibles. I mean, I understand your point, but we have a big job to do."

"There will be times when I'll need you two, but it won't be for long. To succeed, we need to work as a group, to be aligned. So, once in a while, I may call you away from your work and ask for your cooperation in keeping things running smoothly and productively. Likewise, if you have trouble with the plant I would expect you would call the rest of us in to trouble-shoot, brainstorm, move equipment, or whatever it takes to succeed. Maybe we won't have to meet often; maybe we will."

"I agree," I said. "What we are to do at Mars is bigger than all of us. We will need to cooperate. We need to act as individuals but we'll also need to act for the highest good of the crew and of the people of Earth, as best we can. We made a commitment to be a unit when we started. Now we get to make our oneness real. We *are* one."

"What we're doing," Lee said, "may sound like mutiny, like subterfuge. Busby will continue to think he is commander, but I am in charge in a crisis situation. The five of us are highly motivated professionals setting out to do some tasks; we're not deck hands or officers on the *Bounty*. But we're also human beings with emotions and imperfections, and we saw what happened early on the trip. We were all demoralized and unmotivated.

We can't let that happen at Mars. We can't let John get at us. At the same time we can't storm the captain's quarters and drastically take over the ship, send John away somewhere. We can't demoralize him. We *can* quietly support him and at the same time make it clear we are to go on with our work. We can't—"

She stopped abruptly. Busby stood in the doorway of the library. I don't know how long he had been there. We all noticed him at the same instant.

"I'm in command of this ship," Busby said. "And let none of you forget it. But I'm not feeling too good right now. It must be the spinning or something. I do want you to get your tasks done at Mars, I know I can't do it all alone, and so, for the time being I'm releasing the command of this ship to you, Marla. Good night." He walked away.

I had a sudden impulse to express to the group our greater purpose, our oneness in this mission, but knowing the diversity of philosophies and nationalities in the group, I didn't know how to do it. I tried anyway.

"May we do something that might help all of us?" I asked. "We need to do something to feel the oneness of our group. Let's all join hands, close our eyes, and relax for a minute. Remember we did this in one of our training workshops, and it really helped. OK. Let's all breathe in the positive, the beauty of what's here, and the nobility of our task. Now breathe out the negative, the stress, and the feeling of separation. Breathe in again and once again out. May we be filled, surrounded, and protected by the greater energy that is our essence, and may that energy be sent to John Busby, the crew of *Mars 1*, and the inhabitants of planet Earth for the highest good of all concerned. May we approach our tasks and our fellow humans with enthusiasm and compassion."

The group had become calm. We sensed the peace and unity among us. During all those months of training I had endeavored to play down my conviction that dilemmas can be solved through focusing on human potential. Now I realized how appropriate it was to verbalize that ideal and how effective my words had been. A sense of group balance was restored, and at least for a few days, even Busby's vitality renewed.

During November 1998 the red dot of Mars became a disc that grew larger and larger each day. We were ready to swing into action. With the shipboard optical telescope now working, we were getting good views of the planet. On November 30, with Mars now looking as big as the Moon does from Earth, we prepared to despin and reel in the hab and lab modules to their stowed positions behind the aeroshield. We would be weightless for two days prior to aerobraking at Mars. If systems were not operating properly or if we went off course from our precise aerobraking corridor, we would blast the engines and swing around Mars with a free return to Earth. That would be disappointing, I reflected, after all our effort to get here.

The aerobraking maneuver at Mars would take only a few minutes with a maximum loading of four g's. Alongside our spacecraft, just 500 meters away, was the PhD craft. It had its own aeroshield plus its cargo of a nearly empty ET *Ironhorse,* the PhD processing plant, PhD lab, and the Mars vehicle. We had already lost visual contact with *Mars 1,* which was on its own trajectory that would cause it to brake propulsively slightly above the Martian atmosphere, go into a highly elongated elliptical orbit around Mars, and keep braking at each pass close to Mars until it would rendezvous with Phobos and us a few days later.

Meanwhile, we had long since said goodbye to the fleet of unmanned Mars spacecraft that had resembled Multiple Reentry Vehicles (MRVs) at the front of the *Olive Branch* when we first escaped Earth. These rocket-encased Mars landing rovers, balloons, ascent stages with sample return modules, and Mars orbital communications and remote sensing satellites all shot off on their own trajectories toward their predefined places on

Balloons like this one take meterological data on the Martian atmosphere during the Mars 1999 mission.

This unmanned rover is traversing the Martian desert during a dust storm.

Mars and in orbit. In all, there were eight landers, each with a balloon and rover, three with ascent stages for sample returns to Phobos. Also, three Mars communication/observer satellites went into strategic orbits planned for continuous communications with, and observation of, the landing sites. Included in the communication link would be Phobos, the *Olive Branch*, and Earth. The satellites would also continuously survey the planet to monitor its weather conditions. My job was to remotely control these devices—no small task. Quite remarkably, all eleven spacecraft had been working well up to now. The Soviets had three lander-rover–sample return vehicles of their own.

As you can probably see, we had our work cut out for us at Mars. With Busby probably out of the loop, I would be doing most of the rover controlling, with help from Lee, Osho, and the *Mars I* crew—when they could. The Steinmullers would be totally occupied with deploying and operating the PhD plant. Their job was higher priority than mine. The manned Mars landing and economical returns to PhD would rely on quick deployment and operation of the plant. On the first few days all of us would focus our efforts on getting that working.

Unfortunately, we could not see the surface of Mars coming into the aerobrake; the heat shield that had protected us from the Sun was now oriented to protect us from the searing heat of friction with the upper Martian atmosphere. But what happened after the aerobraking was awesome. As we slipped back out of the atmosphere, we could look down a mere 100 kilometers to a huge, red, pristine landscape that was incredibly beautiful. After such a long voyage, the impact of the view exceeded my wildest expectations. Mars seemed to be an Earth rechoreographed in red.

We took our four g's with some nervousness and excitement. Had we stowed our equipment well enough to withstand the coming tremors? Critical parts of the *Olive Branch* had taken ten g's during tests on the Earth, but this was still a first-time event. As the g's began to mount and we mashed into our couches as if large weights were placed on our chests, and as we began to buffet (much like in the movie *2010* at Jupiter), I fleetingly envied the Soviets. They had done a simpler propulsive maneu-

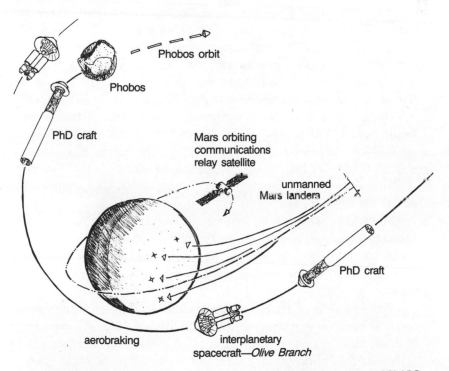

Storyboard of the various spacecraft arriving at Mars and PhD. *Drawing courtesy SCI-ARC.*

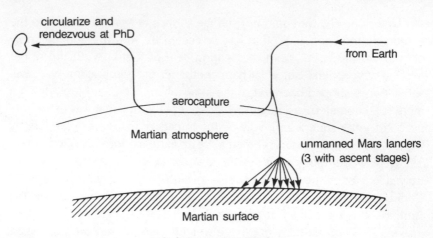

Arrival at Mars (T ± 300 days).

ver above the atmosphere. But we would have had to carry three times more propellant *not* to aerobrake the *Olive Branch*. The Russians had brought along almost as much equipment as we did and they had the big booster. For us, aerobraking was definitely worth it. A mere two hours after our dramatic encounter with Mars, we would be firing our engines near Phobos for a two-month rendezvous with that new world. What a trip!

During all this we were concerned about whether the *Ironhorse* would make it. We thought we had the attitude control engines fixed after the PhD craft tumbled out of control, but we now had to rely on backup engines that would have to work at Mars and PhD.

They did. As we had done, the *Ironhorse* did a perfect aerobrake. We all had the right attitude.

The firing sequence at Phobos was also right on target. Again, until we had actually arrived near there, we couldn't see the moonlet, now sitting about 100 kilometers to our side. The spacecraft attitudes were perfect from a mission-control perspective, but not from an aesthetic point of view.

The first view of Phobos was awesome. In spite of all the pictures taken of the satellite, nothing could match the real thing. We saw a profile of it, the long axis pointing toward Mars. It looked like a big stark black-and-white rock pocked with craters and striations. It was almost twice as long as it was wide. We were now in a similar orbit and would be closing in to a final position about ten kilometers from its surface. The PhD craft would fly alongside us and eventually land on Phobos. We would be sending people to and from Phobos to set up the PhD plant and laboratory.

It is difficult to describe the excitement we felt as we gently closed in on Phobos. Our systems had worked well, but the biggest challenge lay ahead. The *Olive Branch* once again unfurled its two spinning modules, and I took my position at the Mars rover monitoring station and communications center inside the lab module. At the time our near-Mars operations began, only one rover was not functioning. Eventually we fixed it, a temporary problem in the lander's telemetry. All three communications and observation satellites were working.

In all, we would be monitoring eight sites on Mars. Because of the closeness of Phobos to Mars, the data stream of scientific information from the rovers was 100 million times more intense than what could be sent to Earth. This, plus the fact I could "drive" the rovers in real time, not with ten minutes' delay, meant that we could get a huge scientific bonanza while exploring the steep, rocky, sometimes hazardous Martian terrain. We

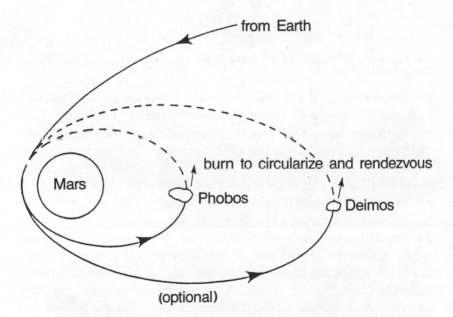

Aerocapture at Mars and rendezvous with Phobos.

View of the *Mars 1 (top)* and hab module of the *Olive Branch* hovering over Phobos. *Drawing courtesy SCI-ARC.*

made the rovers small and produced more of them—eight in all this time around and another eight on the next mission later in 1999.

The eight chosen sites involved an enormous variety of terrains and had the potential for signs of life. One was in Valles Marineris, in Chasma Candor, near the equator, where we would be landing humans in two months if all went well. A second site was near the lip of the giant Olympus Mons on the Tharsis Ridge to the west of Valles Marineris. A third site was in Cydonia at latitude 41 north, a plain with strange features including the face on Mars. That rover would land just to the west of the face and would explore the face as well as nearby pyramid-shaped objects. The fourth site was by the Viking 1 lander site, the Thomas Mutch memorial station, in Chryse Planitia. A fifth site lay at the mouth of one of the dried-up river beds in Mangala Valles. The sixth site was at Chasma Boreale near the Martian north pole where it was now summer and bathed in constant

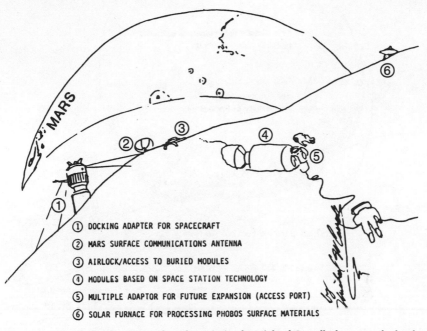

① DOCKING ADAPTER FOR SPACECRAFT
② MARS SURFACE COMMUNICATIONS ANTENNA
③ AIRLOCK/ACCESS TO BURIED MODULES
④ MODULES BASED ON SPACE STATION TECHNOLOGY
⑤ MULTIPLE ADAPTOR FOR FUTURE EXPANSION (ACCESS PORT)
⑥ SOLAR FURNACE FOR PROCESSING PHOBOS SURFACE MATERIALS

Diagram of the Phobos base. *Based on the painting by Michael Carroll (shown on the book jacket).*

(but cold) sunlight. The seventh site was on the fringes of the maximum extent of the south polar cap with its peculiar layered deposits and fretted terrain. And the eighth site lay within the plains of Elysium near its peculiar triangular-shaped pyramids.

The scientific studies we would be doing in the next sixty days would cover more areas of Mars in more detail, by factors of millions, than one or two Earth-controlled rovers could. The rovers were my job, a responsibility I found both exhilarating and unnerving. Fortunately the *Mars 1* crew included Claude Michel, a French geologist and archeologist who would be helping me make some of the major on-the-spot decisions. There was plenty of help from Earth, too, but the main job was mine.

My control station on the *Olive Branch* included eight television monitors, one for each site, to provide continuous, simultaneous coverage of rover views of the traverses around the sites. It was like a television program director simultaneously looking at various camera views and taking appropriate action. Meanwhile, data readouts on the meteorologi-

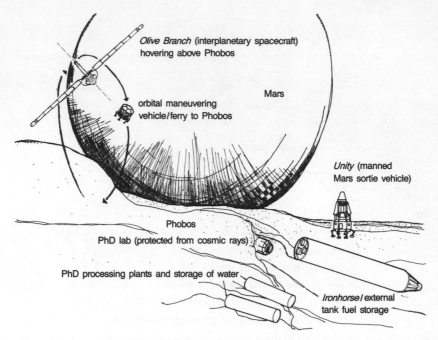

Operations on Phobos. *Drawing courtesy SCI-ARC.*

cal conditions at all sites would be continuously streaming in, also from balloons that were following the prevailing winds around the planet.

The rovers operated something like the Soviet Lunakhod Earth-controlled lunar rover back in the 1970s. I would "drive" them (sometimes several simultaneously), occasionally stopping them to pick up an appropriate sample that could be thrown into the hopper and returned to the lander for analysis. Rovers at the three prime sites—in Valles Marineris, Cydonia, and Mangala Valles—had ascent stages that would deliver samples to the PhD lab in about a month, so they could be screened for harmful contamination and later returned to Earth for detailed analysis.

Simply put, I was *busy.* My working days were sixteen hours, sometimes more. Osho, Lee, Michel, and even Busby worked with me and filled in for me while I slept.

The real excitement began in the second week of rover teleoperating. The Site 3 rover, approaching the base of the "face" in Cydonia, happened upon some strange angular protrusions that stuck up several tens of centimeters above the sandy surface. The rover chemical analysis showed an aluminum and titanium content with a weathered, iron oxide coating, like the rest of Mars. What was it? Normal attempts to expunge it from the surface failed, and so we ran over it with the traction wheels, trying to pry it loose. Still no luck. Were these possible artifacts of an ancient civilization? Further traverses showed more evidence that they were.

Meanwhile, on the surface of Phobos, the Steinmullers were overseeing the PhD plant. Setting it up required a lot of EVA, a lot of care, and a lot of help. Osho was with them most of the time. The first step was to make sure all PhD landed modules were intact, anchored to the surface, and ready to operate. The huge *Ironhorse* lay on its side just barely around the limb of Phobos out of line of sight with Mars. This kept the temperatures there to a minimum for storing liquid oxygen and liquid hydrogen. Just 100 meters away on a rise at the limb were the two mining tractors, encased in a large, slightly pressurized, inflated Mylar dome (which kept dust from spreading), the solar furnace that would heat the Phobos dust, and the PhD lab, telescope, and control center. Not too far away were the radio communications dishes, the Mars sortie vehicle, and water storage tanks, which used to be logistics modules on the *Olive Branch*.

It was important to fasten each of these modules onto Phobos so they wouldn't float away. The gravity on Phobos is one-thousandth that on Earth, so working there was similar to working on the space station. The mining operation under the Mylar bag was small and environmentally safe.

The plant began to work like a charm on the third day of our stay. The processing was simple. A small tractor pushed Phobos dust onto a small conveyor, which moved about one kilogram of the dust every three seconds into the focus of the solar furnace. The lightweight furnace consisted of a steerable, aluminum-coated, gossamer structure about fifty meters in diameter. It had been erected in a matter of minutes. Sunlight concentrated onto the focus of the mirror produced about three tons of water per day. We stored two tons in tanks, and the third ton went into a tank containing various chemicals. This portion of the water once again passed through the focus of the solar mirror. This thermochemical process produced oxygen and hydrogen, which would be liquified and transferred to the tanks of the *Unity* (the Mars Excursion Module) and the *Ironhorse* (the ET).

Between December 18 and January 15, we analyzed Martian samples collected by the rovers from three sites on Mars aboard our small laboratory on Phobos. We found that the samples were not contaminated with Martian biota, so the coast was clear to send people to Mars. We had also analyzed Phobos and Deimos samples and found them to be similar to the carbonaceous chondrite meteorites that land on Earth. The soil was full of water, hydrocarbons, and amino acids. But because Phobos has been untouched by Earth's atmosphere, we were for the first time directly sampling the primordial matter of the solar system and were bound to find out a lot more about where we came from.

With everything working in synchronization, we were ready for the January 27, 1999, sortie to Mars. Time flew so fast, Christmas and New Year's Day were but fleeting celebrations in the most extraordinary human quest of all time.

We found signs of possible life on Mars, we landed people there, we directly sampled the primordial material of Phobos and the red sands of Mars, we resupplied our water and fuel to land on Mars, and soon we would be creating and stockpiling thousands of tons of water and sending it in many directions. We had reached humanity's great turning point in space. Beyond it lay our galaxy and the infinite vastness of the universe. Humanity had turned its spiral outward to the stars.

8

Why Phobos?

Phobos and Deimos are the two odd, rocky, potato-shaped moons of Mars. Phobos measures twenty-seven kilometers and Deimos fifteen along their longest axes. By Earth's lunar standards, they are more like asteroids. In 1976, the U.S. Viking orbiter photographed both moonlets, showing surfaces of irregular lumpiness pocked with millions of craters from the impact of meteoroids over the eons. The smallest specks of craters measure barely fifty meters across, about one-half the length of a football field. Yet a chunk of either moon fifty meters wide could probably supply tens of thousands of metric tons of water and fuel to take care of transportation and life support for growing bases on Mars, the Moon, and space settlements.

The study of Phobos and Deimos has had an interesting history. Using mathematical reasoning that had no relation to physical reality, the German mathematician, astronomer, and mystic, Johannes Kepler, predicted in 1610 that Mars has two moons. The writings of Swift and Voltaire during the following century also mentioned the two moons of Mars. Yet nothing had been observed.

Viking 2 photograph of Phobos. The tiny moon's north pole is near upper left. Phobos is heavily cratered, as expected, but, surprisingly, shows striations and chains of small craters.

Not until 1877 did the Martian moons materialize in telescopic observation. Asaph Hall at the U.S. Naval Observatory noticed two faint starlike specks moving in orbits around Mars. He named the satellites Phobos and Deimos after the two sons of the Greek god Ares (whose Roman name was Mars). Phobos means fear and Deimos means terror, not very attractive thoughts for would-be voyagers.

Their names notwithstanding, Phobos and Deimos are tiny moons, barely observable as dots through large telescopes. These asteroidlike objects may look strange in comparison to larger spherical bodies like Earth, Mars, the Moon, and the Galilean satellites of Jupiter. Astronomers have observed thousands of asteroids (*aster* is the Greek word for star, or point of light) that inhabit a belt between the orbits of Mars and Jupiter. Asteroids vary in size from a few kilometers across to very large objects like Ceres, about 1,070 kilometers in diameter. Like Phobos and Deimos, many appear to contain a wealth of material resources. Only a tiny speck of Phobos or Deimos is all that will be needed to supply a growing space industrial infrastructure; yet neither moon is large enough for the strength of its mate-

A Viking 1 photograph of Deimos.

rials to succumb to gravity and become spherical like other moons and planets in our solar system. The material resource available to us is incredibly vast.

Phobos and Deimos might be captured asteroids. Yet their almost perfectly circular orbits around Mars at the equator, a little less than one Mars diameter away for Phobos and three for Deimos, make it difficult for scientists to explain how they could have gotten there from the asteroid belt. But it is certainly possible that the moonlets slowly evolved into their present orbits.

Within a relatively short time in the life span of Phobos and of the solar system—less than 100 million years out of 4 billion—Phobos will crash into the surface of Mars, unless someone engineers a solution. Tidal drag, or the

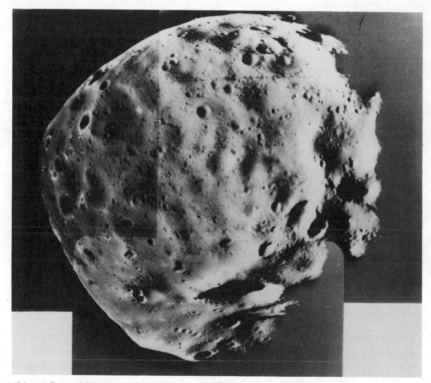

Viking 1 flew within 300 miles of Phobos to get the pictures in this mosaic of the asteroid-size moon. North is at the top. The south pole is within the large crater, Hall, at the bottom center where the pictures overlap. Remarkable features include striations, crater chains, a linear ridge, and small positive features that appear to be resting on the surface.

differential pull of Mars on the front and back sides of Phobos, is causing the moon to slowly spiral in toward Mars. We are living at a time when Phobos will be first visited by humans, just in the twilight of its life.

Both moons are very dark, reflecting only about 5 percent of incident sunlight. Their average densities, as measured by their gravitational pull on each other and on space vehicles, are very low, about two grams per cubic centimeter. This is twice the density of water but only two-thirds the density of common silicate rocks found on Earth, the Moon, and rocky meteorites. This information suggests strongly to scientists that Phobos and Deimos are composed of materials similar to what we find in carbonaceous chondrite meteorites that land on Earth. Most carbonaceous chondrites contain large quantities of water (10 to 20 percent by weight) bound in the soil and varying amounts of carbon compounds. When they are analyzed, we sometimes find they contain hydrocarbons, sticky tars and oils, and even amino acids. Scientists believe carbonaceous objects comprised the primordial matter of the solar system before planets formed. Directly sampling these pristine materials, unadulterated by earthly processes that have affected meteorites, would be a big step in piecing together how the solar system formed and how we got here. Phobos and Deimos are great candidates for providing that step, another bonus of Mars 1999.

The "water of hydration" absorption band, found in the infrared spectrum of some large asteroids, indicates that many asteroids are also carbonaceous and contain more than a few percent water by weight (depending on the intensity of the band). They are also very dark objects, which reflect very little of the incident sunlight. From this small data base, scientists are able to classify asteroidal compositions by measuring their reflectivities and colors. A large number (roughly half) of all asteroids are classified as carbonaceous or "C" objects. Phobos and Deimos are "C" objects, strongly suggesting the presence of water in their soil.

The bad news is, we don't know for sure whether water is on PhD. They are too faint for us to detect the water of hydration band from Earth-based telescopes; and no spacecraft has yet been equipped to observe it. The evidence is circumstantial, but it very strongly suggests water is indeed present on PhD.

The good news is, we will probably get an answer soon. In 1988, the Soviets plan to launch two unmanned spacecraft to Phobos. One of these may also go to Deimos. Both spacecraft will arrive at Phobos in early 1989. The first one will swoop down and zap a square millimeter of the upper-

This Viking 1 photograph shows Phobos (black spot) passing beneath the spacecraft with the surface of Mars in the background. One of the darkest objects in the solar system, Phobos is four times darker than Mars. Phobos appears black because the exposure in the camera was set for the Martian terrain.

most layer of soil with a laser. As the dust evaporates, a special instrument will determine the composition of the particles thrown out.

The Soviet Project Phobos also provides for a small long-life autonomous station on the moonlet. When one of the spacecraft nears Phobos, the station will separate from the space vehicle, spin for an oriented landing, and become attached to the surface with a probe driven into the ground by small explosives. The lander will be equipped with a television camera and instruments that will chemically analyze the soil as well as its temperature and physical characteristics.

A second lander robot will be able to make similar measurements at several sites over the surface of Phobos by hopping in spurts of twenty meters or more. The low gravity on Phobos, one one-thousandth of Earth's gravity, permits unique designs of this sort and makes exploration mechanically easier than in the presence of either no gravity or substantial gravity.

Because of the redundancy of systems, it is highly likely that the Soviets will provide the world with a final answer on whether water is present. We will also find out how easy it will be to scoop up the soil for its processing into water. The Russians have invited Americans to participate in Project Phobos. Things are definitely loosening up, and the Soviets are also gearing for success. The reason that Phobos is a higher priority for them should be obvious: they too find Mars 1999 a desirable scenario.

What would life on Phobos and Deimos be like? The low gravities of both moons raise all sorts of unique possibilities. The escape velocity from Phobos is a mere forty kilometers per hour; from Deimos, twenty-five kilometers per hour. These are the speeds of a car on city streets, a running dog, a baseball in playing catch—not the Mach 35 we need to get off Earth or the equivalent of fast bullets required to get off Mars and the Moon. Orbital velocities around Phobos and Deimos close to their surfaces would be about thirty and twenty kilometers per hour, respectively. Could you imagine standing, floating, or hopping near the top of a rise, pitching a baseball horizontally, and catching it coming in from the opposite direction (or having it hit the back of your head) about three hours later? You could launch your own satellite and retrieve it all by yourself.

Keeping the same faces toward Mars, both moons make ideal platforms from which to view the Martian surface. There would be no need to stabilize orbiting satellites on three axes and worry about how long they would last as they gradually run out of attitude-control propellants. Deimos, with its thirty-hour revolution period around Mars, is just outside Mars synchronous orbit, where the period of satellite revolution would just equal the period of one Mars day, or twenty-four hours, thirty-seven minutes. This means that, seen from a site on the Martian surface, Deimos will rise in the east and slowly move across the sky and set in the west two days later. Phobos, on the other hand, rises in the west, whizzes across the sky in less than six hours, and sets in the east. Phobos would appear in the Martian sky as a peculiar potato, a little less than one-half the size of our Moon as seen from Earth. Deimos would appear to be six times smaller and be fuzzy.

The first man to point out some of the unique scientific attributes of Phobos and Deimos was Fred Singer, a physicist from George Mason University and one of the most innovative thinkers of our time. In 1968, Singer proposed that the quality of science performed by rovers on the Martian surface and sample returns to Phobos or Deimos could be significantly enhanced by having astronauts on one of the moons controlling the rovers and screening returned Martian samples for their possible contamination. It was Singer who coined the abbreviation PhD for these moonlets. His ideas have since been expanded into the more holistic Mars 1999 concept. He wrote in the October 6, 1986 issue of *Newsweek:*

> Technically, such an operation means little more than transferring to Mars orbit a space station similar to the one presently planned for close Earth orbit. . . . From the standpoint of science, the use of Phobos and Deimos as bases would be more rewarding than a manned landing on Mars. Because of the planet's terrain, humans wouldn't be able to venture far, nor could they operate a robot vehicle without the use of a satellite since Mars's mountains would block their view. They might as well be exploring Mars from Deimos. Using Mars's moons as a base would also be better than unmanned exploration directed from the Houston space center. Because of the distance, radio signals to and from Mars can take as long as an hour. And "driving" an unmanned rover from Earth, step by step, can become a time-consuming operation. Sample returns to Earth would take months instead of hours, follow-on missions would be years apart instead of days, further slowing the process of exploration.

Singer argues convincingly that time and money are saved by exploring Mars from PhD, compared to doing it either unmanned from Earth or manned on the Martian surface. The contamination problem can be handled much more easily too, because we could analyze the appropriate samples and select safe landing sites on the spot rather than performing a long sequence of round trips from Earth.

The rates of return of data from the Martian surface to a receiving antenna on Phobos or Deimos would be *10 million to 100 million times greater* than the data return to the same antenna stationed on Earth! Even with a much smaller antenna on PhD, we could be getting a million times more data. From PhD, we can explore Mars in real time, using television transmissions as seen through the eyes of the rover cameras, in much the same way that the Russians in the 1970s explored the lunar surface with the Lunakhod rover. And we can use several rovers at several sites at far lower cost than any other proposed scenario.

During the late 1970s I became intrigued with Singer's ideas. At the time I was working with Gerard O'Neill at Princeton University on concepts of using nonterrestrial materials as a basis for industrializing and colonizing space. My findings pointed toward Earth-approaching asteroids as good early destinations for mining raw materials. I discovered that some of these objects at certain opportunities were more accessible to Earth than our own Moon in terms of propulsion and expense. It was also clear that the asteroids offered a greater variety of raw materials than the Moon could supply.

From meteorite analysis and spectral observation through telescopes, we knew some asteroids were loaded with water and carbon compounds while others contained an abundance of free metals. Some meteorites have higher enrichments of platinum group metals than the richest ores on Earth. We found that, ultimately, any material we wanted to use in space would become available to us from the asteroids at a small fraction of the cost of sending it from the deep gravity well of Earth. Eventually all we would need from our home planet would be seeds and people! I discovered that the material resources are vast. All we would need is less than one-billionth of available asteroidal resources to extensively industrialize and colonize space.

As interesting as the asteroid ideas were, it seemed that there was a missing element. In these times of conservative thinking, the idea of mining asteroids just seemed too far out to enter into serious space planning. Humanity's grand debut in space needed a nearer-term, more feasible-sounding set of events.

Then a wonderful thing happened, one of those elegant accidents of science. I found out that, as accessible as some of the asteroids are to Earth, Phobos and Deimos are just as accessible when we consider aerobraking at Mars and Earth. Not only that, PhD are more accessible *more often* than any other known natural objects in the solar system. Launch windows to the Martian moons occur every two years, whereas the good launch windows to the best-situated Earth-approaching asteroids occur at intervals of one or two decades. Whereas missions to mine a given asteroid would be rare, we could keep cycling spacecraft and materials to and from PhD. Even without aerobraking, PhD are easier to go to and come from than our own Moon; using the Moon for a gravity assist coming back to Earth from PhD makes the difference even larger. These considerations become very important when we begin to talk about transporting large tonnages of materials.

It was pure serendipity. To my way of thinking during the late 1970s, Phobos and Deimos appeared to be the most accessible objects to Earth known to humankind. They were the pick of the asteroid litter. Plus, they probably contained water and hydrocarbons, shown in engineering studies to be by far the most useful and economical materials for the buildup of space industries. Add to that Singer's convincing arguments for exploring Mars, and then the political groundswell toward Mars, and we have a certain winner.

Where do we go first, Phobos or Deimos? Probably the most important consideration in making that decision will be their chemical compositions: which moon has more water of hydration contained in its surface soil? We hope to get an answer in 1989 from the Soviet Phobos mission, which may also visit Deimos. If the Soviets fail or if water is not found in Phobos, a U.S. precursor mission will be necessary, or perhaps a reprogramming of the 1992 Mars Observer mission. If the Soviets find plenty of water on Phobos and don't visit Deimos in 1989, we might choose to go to Phobos by default. If it has what we need, why worry?

In the unlikely event water is not found on either moon, oxygen certainly will be found in the form of chemically bound oxides. Oxygen, constituting 73 percent or more by weight of a propellant mixture with hydrogen, would be worth processing. We would need to design and send an oxygen processing plant similar to ones now being designed for the lunar surface. In such a case we would probably bring hydrogen from Earth. It would also take five to ten times longer to process the oxygen from the oxide state than it would from water because of the greater bonding energy of oxides.

Let us assume that the composition of the two moons is similar. Where do we go first? As of this writing I don't know; there are advantages and disadvantages to each satellite. While Phobos is closer to Mars, affording higher-resolution views, Deimos is in a nearly Mars-synchronous orbit, allowing longer continuous coverage of a given site on Mars (days versus hours). Deimos provides an advantage in the event that tight budgets prevent a comprehensive network of Mars communications and remote sensing satellites. Phobos has twice the surface gravity of Deimos, and this might make surface operations easier.

Deimos provides a better vantage point from which to view higher-latitude Martian sites (up to 81 degrees latitude versus 68 degrees for Phobos). However, Phobos has more access to the surface of Mars near the equator. It takes 2.2 hours to get to Mars from Phobos and 6.7 hours from Deimos. We would need less propulsion to travel between regions at the Martian equator and Phobos. The time difference may be appreciable if we require that the manned Mars landing sortie during Mars 1999 be a sprint because of limitations on the crew life support.

On the other hand, in its position farther out of the Martian gravity well, Deimos is closer to escape to trans-Earth injection. But for Mars aerocapture, the propulsion needed to rendezvous with Phobos is slightly less than that at Deimos. If we use liquid oxygen and liquid hydrogen propellants, these very cold fluids would boil away more rapidly near Phobos than near Deimos, unless the storage tanks are on their back sides, facing away from Mars. On the other hand, a telecommunications and remote sensing base on PhD is best placed on the Mars-facing sides, suggesting two sites per moon.

It will be easier to change orbital planes at Deimos than at Phobos, with more accessibility to higher latitude sites. However, the propulsion needs in both cases are sufficiently high to argue for manned sorties to sites near the equator from both moons, and Phobos is better for that.

Fred Singer favors Deimos; other scientists prefer Phobos. Of one thing we can be certain: Mars 1999 will establish its base on one of the moons and a sampling sortie will probably be made to the other moon. If I had to make a choice now it would be Phobos, because that is the Soviets' main objective. Phobos is where we are bound to find answers in 1989 about whether or not there is water.

What would a water and fuel processing plant on PhD look like? Engineers and scientists have looked at many candidates, with no final decisions yet. The technology is simple, and the projected productivity rates are high. Testing and deployment will present a formidable challenge to designers and operators, but there are certainly no surprises in store.

The basic processing will take two steps: (1) extract water from the soil; and (2) separate some of that water into oxygen and fuel (hydrogen, methane, or propane). The first step is easy unless the soil is not loose. To take a worst-case situation—solid rock—we would need drills, crushers, blowers, dust containers, pipes, tractors, and conveyors. It is far more likely that the soil will have the consistency of fine beach sand and could be easily moved to the plant by a small tractor and conveyor belt. The Viking photos of PhD show moonlike surfaces exposed to the constant barrage of micrometeoroids through the ages, where the surface layers have been pulverized and gardened into a fine dust. Carbonaceous material is loosely agglomerated and would be even easier to manage than lunar dust.

We would then pass the PhD dust through the focus of a solar furnace, driving out the water vapor and other volatile materials such as carbon compounds. The products would flow through pipes into storage tanks, where they would condense into water, ice, dry ice, and so forth. The solar furnace would be a steerable aluminized Mylar dish, deployed to about fifty meters in diameter. The shape is parabolic to focus the sunlight onto the PhD material. Such a flimsy structure that could steer to face the Sun yet be moored to a solid surface could not be built on Earth, the Moon, Mars, or even in outer space; PhD is the perfect setting for such a solar furnace.

Assuming 10 percent of the PhD soil is water and a goal of producing three metric tons per day of water, we would need to pass about one pound of PhD soil into the focus of the furnace every ten seconds, to be heated up to about 300 degrees Celsius. A fifty-meter-diameter furnace would be more than adequate to do the job. Because a given site on PhD is sometimes in nighttime or in Mars eclipse, we will not be able to use the furnace all the time. The problem could be minimized by placing the reflector on a high point. An alternative would be to use a nuclear reactor that could produce both the needed thermal energy and electricity. However, nuclear systems require shielding and have a bad reputation, for right or wrong.

The second step is to convert some of the water to propellants. The most direct route is to produce liquid oxygen (the oxidizer) and liquid hydrogen (the fuel). Rockets using these substances have the highest efficiency. The main problem is the need to refrigerate them and to keep them from boiling away as they are stored in tanks. Using insulated tanks within the shuttle external tanks might be one solution. Another possibility is to process for methane or propane as fuel and nitrogen tetroxide or carbon monoxide as the oxidizer. All these possibilities need to be looked at.

Assuming we go the simpler route to liquid oxygen and liquid hydrogen, what do we need to send to PhD to make that happen? At least two methods are possible and there might be others. The first is electrolysis, in which water is divided into hydrogen and oxygen by electrical means. This process is well understood and, interestingly enough, has been proposed for use in Earth orbit.

During the 1970s, engineers Ed Bock and J. G. Fisher of General Dynamics proposed an electrolysis system in orbit that would receive water launched by the space shuttle and then process the water into liquid oxygen and liquid hydrogen. The storage tank would be a specially equipped shuttle external tank that would be used later as a rocket to boost payloads toward higher destinations. The idea has been recently revived as new, stringent safety and weight requirements limit what goes into the shuttle cargo bay. Rockets that will send cargoes and people beyond low Earth orbits (called Orbital Transfer Vehicles, of which the current model is the Centaur) may need to be launched from Earth without propellants. We would need to fuel them in space. Scientists Peter Vajk and Philip Chapman have proposed an Earth orbital electrolysis system to convert shuttle-launched water to propellants for Centaurs.

A second system, proposed by the space engineering firm JDR Associates, would produce the liquid oxygen and liquid hydrogen from water using chemical and thermal processes and a solar furnace. On PhD, if we have a solar furnace (or nuclear furnace) already in place to extract water

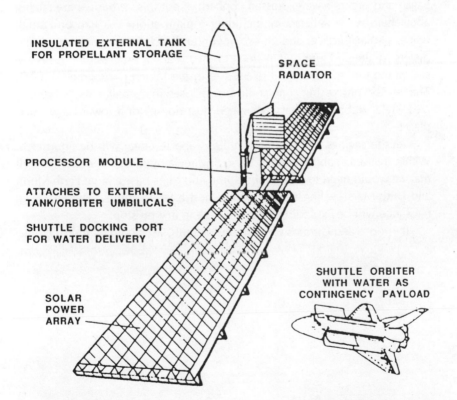

INSULATED EXTERNAL TANK
FOR PROPELLANT STORAGE

SPACE
RADIATOR

PROCESSOR MODULE

ATTACHES TO EXTERNAL
TANK/ORBITER UMBILICALS

SHUTTLE DOCKING PORT
FOR WATER DELIVERY

SOLAR
POWER
ARRAY

SHUTTLE ORBITER
WITH WATER AS
CONTINGENCY PAYLOAD

Concept for LEO water processing into LOX/LH$_2$ to be stored in an ET with application to PhD water processing.

from the soil, it makes sense to use the same furnace to thermochemically reduce the water to liquid oxygen and liquid hydrogen.

Interestingly, both the electrolysis and thermochemical systems result in the production of about thirty metric tons of propellant per month. This is almost exactly what we would need at PhD to make the human Mars sortie thirty to sixty days after arriving at PhD. It is also what we would need to fill the external tank over a period of two years—in time to return the first payload of water to Earth orbit at the next PhD departure window during January 2001. The total system would weigh about twenty metric tons and therefore be able to produce its weight in water in one week and in propellants in three weeks. We see a lot of serendipity in the commonality of these systems, an important economic consideration in NASA planning.

The PhD processing plant needn't be an environmental disaster. Carl Sagan and others have expressed concern about this. What we are talking about here is a small facility including a plant about the size of a small office, a solar furnace, and an excavation and reclamation area about fifty meters square and four meters deep for two years of processing. This is the size of the tiny speck crater I referred to in the Viking photograph of PhD. The "messy" processing zone could be enclosed in a small, slightly pressurized Mylar bubble, so that high-speed dust doesn't drift toward unwanted places.

Yet the savings will be in the billions and leverage will be enormous. Within five years, about eight of these tiny, reclaimed strip mines will be all that we would need to create large lunar and Mars bases, large Earth-Moon and Earth-Mars cycling space stations, and asteroid mines. The space renaissance will be exploding, with PhD as the first pit stop.

If Phobos and Deimos were moons of Earth, we would be there by now.

9

Scenario

New Year's Eve 1999

• JANUARY 25, 1999— A second U.S. manned spacecraft, Enterprise, and its crew escape Earth orbit on a trajectory toward Mars. • JANUARY 27—One Soviet cosmonaut from Mars 1 and one U.S. astronaut from the Olive Branch make the sortie in the Unity from Phobos to the surface of Mars and return to Phobos. • JANUARY 31—Both U.S. and Soviet crews aboard the Olive Branch and Mars 1 launch from Phobos for the journey back to Earth. The fuel-processing plant on Phobos continues to stockpile fuel. • SEPTEMBER 9—The Enterprise and crew arrive at Phobos and establish a second fuel plant. • NOVEMBER 9—The Olive Branch and Mars 1 and crews land on Earth for the completed mission. • DECEMBER 31—While Earth residents cele-brate the turning of the millennium, the second U.S. crew collects fossils on Mars. Almost complete nuclear disarmament is achieved. • FEBRUARY 29, 2000—The Enterprise crew encounters extraterrestrial intelligence on Mars.

I was nervous, my heart pounding. I poured myself some more pineapple juice and walked out to the lanai overlooking the surf that lapped against the Na Pali cliffs about four miles to the west. Clouds were peeling off Mt. Waialeale and the rainbow to the east was bright now, the sun shining through the rain. Wet tropical aromas filled the air.

Hanalei never looked more beautiful, but somehow I wasn't fully present. Was it the return-to-Earth syndrome, the post-mission debriefing, the exhausting, exhilarating world tour? Or was it a deeper anxiety? Why hadn't the superpowers achieved the complete disarmament initiative we said we would? Why weren't people on Earth willing to be together as one, as we had been on the *Olive Branch?*

The Sun was about to sink into the Pacific. The dirigible would be coming in through the rainbow to the east, riding the trade winds from Honolulu. Marla Lee was coming to visit me. Maybe that's what I was nervous about.

I had known her for years, and the journey to Mars made us even closer. Our mission. . . . Except for the rough start, our engineering and scientific accomplishments were spectacular. We had a processing plant working on Phobos. We had some pretty incredible results from the Mars rovers that would keep me busy for the rest of my life just reporting to other scientists what I had seen: the metallic objects in Cydonia and the blue-green walls of Valles Marineris, and the seabed fossils of Chasma Candor. An unqualified success. Why, then, had I gone into seclusion in Hawaii? Maybe I just needed to catch my breath.

Our mission back to Earth from Phobos had gone smoothly. We had blasted off from Mars orbit on January 31, 1999, and arrived at Earth orbit on November 9, 1999. I reflected that if the date 8-8-88 marked the intention to voyage into infinity, November 9, 1999, would be a good date to finish the mission, if we take into account the fact that November used to be the ninth month of the calendar. The number nine signifies completion in various ancient traditions.

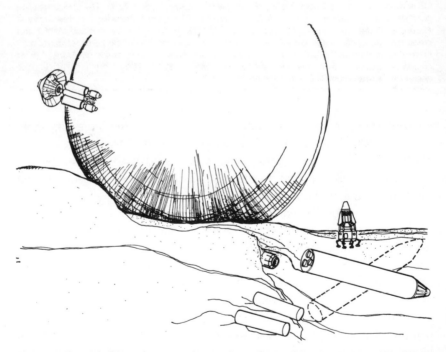

Departure from Phobos. Here, the external tank is reoriented for departure in 2001, after it has stockpiled PhD propellants. *Drawing courtesy SCI-ARC.*

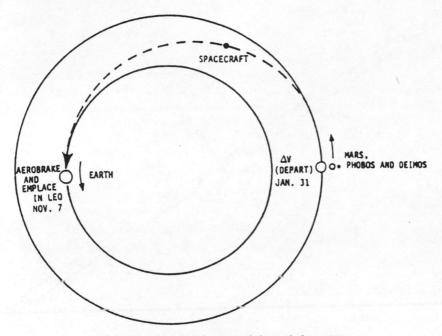

Minimum-energy transfer to Earth from Phobos: 1999.

The return mission had been one of continuing interpretation of the mother lode of scientific data on Mars, Phobos, and Deimos. It was also a time of celebration. Crew exchanges with the Soviets were much more frequent, and I had become conversant in Russian and much more skilled at chess. Two of the *Mars 1* crew members had had difficulty retaining enough bone calcium for good health, and so we invited them to spend some of their time with us in the spinning *Olive Branch*.

Busby had felt better during the return. Although no longer in command, he had settled into a more cooperative frame of mind and helped us with our data reduction. He had seemed proud of our achievements and also happy we were on our way home. And we still had played the games we had invented in variable g.

Part of my anxiety now probably came from the fact that the Soviets had pulled out of their *Mars 2* mission. They had originally planned to fly alongside the *Enterprise*, our second interplanetary craft, but then suddenly canceled the mission. Why? Was it the cooling of relations with the U.S. after their failure to allow the American delegation in to inspect one of their

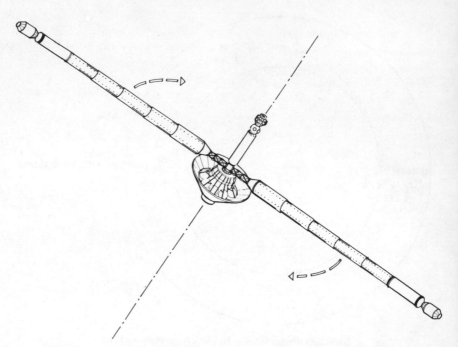

Interplanetary spacecraft on the return voyage. *Drawing courtesy SCI-ARC.*

last nuclear sites? Were they stockpiling warheads while we destroyed ours? Or was it simply their ailing economy?

In January 1999, just before we landed on Mars, the *Enterprise*, with an international crew of seven, had blasted off from Earth orbit for a second human encounter with Mars. The *Enterprise* closely resembled the *Olive Branch*, but the payload was different. Not included were a second PhD lab and second complete Mars Excursion Module; the *Unity's* ascent stage rested on the surface of Phobos, awaiting another sortie to Mars. Instead, the crew of the *Enterprise* carried along a completely equipped space station module that would land on Mars in November and become the home for four crew members for fourteen months.

The *Enterprise* arrived at PhD on September 9, 1999 (nine, nine, ninety-nine), two months before we returned to Earth. A second processing plant at Phobos began stockpiling more propellant for the second landing on Mars and for the return to Earth. The *Enterprise* flew on a three-year round-trip "conjunction class" mission and did not use aerobraking at Mars and Earth.

In January 2001, one year after the big New Year's celebration, the crew of the *Enterprise* would take off from Phobos, with two ET hauls of PhD water fueled by PhD propellant. In late 2001, we would begin cashing in on our investment: 2,500 metric tons of PhD water would be processed in Earth orbit into propellant to send more cargoes and astronauts toward Mars in 2003. Humanity's emergence in space was well under way.

As we were celebrating the new millennium, astronauts at the first Mars base in Chasma Candor were collecting fossils, and a new rover and ascent stage in Cydonia were collecting the metal samples and rocketing them to the PhD lab.

Everything was working out so well. I had gone on the grandest adventure in all of recorded history. Why did I feel so empty?

Marla's visit to me—the first time we really had a chance to get together after all the fanfare—transformed me (once again). During these self-delusionary times, I thought I had learned everything, that I was *the* expert on Mars, that I was a hero. Yes, I had fulfilled my duties, I had enjoyed the journey of the *Olive Branch*, but I *really* wasn't in tune with life on the planet Earth. Marla knew all that about me—my vulnerable spots.

The truth was, here I was back on the planet—gravity was literally pulling me down more strongly than I was used to or wanted it to. I felt helpless because I believed Earth could benefit from my insights, but it resisted. Earth is perfectly imperfect, someone once said.

My doorbell rang, and there stood Commander Marla Lee. We hugged

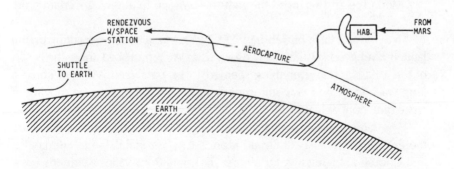

RETURN TO THE EARTH (T ± 650 DAYS)

for a long time. Beneath the commander and the physician was the friend I needed.

"Am I happy to see you!" I said.

"Same here," she replied. "You look a bit pale. I thought you'd be out there soaking up the sun."

"Nah. I'm feeling a little low. I don't know what it is, really. But you, you look great. Come on in."

"I bet you do know what's going on with you. Tell me."

I sighed. "Well, it's just that everything seems such a mess. We've done so many wonderful things in space, but we can't seem to handle our own planet. Do you think the Russians are hiding warheads from us? How about the Bulgarians? The U.S. occupation of Mexico?"

"Come on, you know better than that. Just because we went to another planet doesn't mean we can snap our fingers and change this one to suit our mood. Wasn't it you who told me the world must unfold as it does? Aren't you the one who suggested that I "let go and let God—you know, relax and be patient?"

I didn't say anything, so she continued. "*Use* that great ability of yours. Strive for excellence, even now, even after the big adventure. The real adventure continues. Right? The one *inside*. Remember?"

She smiled, I laughed. What a relief. What a remarkable woman.

I looked at my watch. It was almost seven o'clock in Hawaii. In just a few minutes the next millennium would arrive on the East Coast. I turned on the television set and we watched a frost-huddled crowd await the dropping of the traditional ball from a clock in Times Square.

Marla Lee and I danced through the passage to a new, uncertain time.

Two months later, on February 29, 2000, Marla and I were sitting on the lanai listening to Hawaiian New Age music. We reminisced about the NASA of the 1980s, when everything seemed to be paralyzed. We had come a long way! The world was still imperfect, but we seemed to be moving in *some* direction, at least.

The radio announcer interrupted with a bulletin. "The crew on board the *Enterprise* has just confirmed reports that astronauts Jamie Hernandez and Chou-En Ho, missing for almost six hours from Valles Marineris base, have returned safely to base claiming to have had an encounter with extraterrestrial beings. We have no more details at the moment, but we'll keep you posted." The announcer paused. "If our astronauts weren't just breathing funny air, stay tuned for one heck of a story."

"Brian!" Marla almost shouted as she jumped up and turned off the radio, out of character with her usual even-handed style. "I *knew* these guys," she said. "I'm *sure* their encounter was real."

That was to be the first of many such encounters.

10

NASA, the President, and the People

To some of you the scenario of the last chapter may be long on fantasy and short on credibility. Can we really accomplish a program like Mars 1999? The sad truth is, we won't be able to do it in today's climate. Today's paralysis will be tomorrow's paralysis unless the workings of the institutions and the attitudes of individuals at the helm change toward the positive. The prerequisite to a successful Mars 1999 program is not engineering feasibility. It is people. And there is hope.

Meanwhile, as the dust settles from *Challenger*, NASA continues to search its soul. In the wake of the accident, it becomes all the more evident that the U.S. civilian space program has been suffering from conflicting interests and goals, intercenter rivalries, uneconomical operations, and an apparent inability to make the sweeping changes that are required. Management of the space station program suffers from this confusion. The space agency's technical achievements have been, and continue to be, extraordinary. Nowhere can more intelligent and competent engineers and scientists be found. But there appears to be a bureaucratic inertia that inhibits the innovative thinking and risk taking required to blaze new trails.

The space program needs vital leadership. It needs fresh thinking at all levels, from the President to the project engineers, from astronauts to managers, from advisory committees to public space groups, from entrepreneurs to scientists. Before too long, many of our current leaders will be retiring, and there will be opportunity to make changes that will open up the inevitable human space renaissance in the twenty-first century.

Specifically, what is happening in NASA? Cost overruns, delays, changes in management, and the possible cancellation of the space station, NASA's biggest hope and plum, continue to threaten the very existence of the U.S. space agency. All of us need to be aware of this situation.

The beginnings of what I see as NASA's inability to set bold goals and to achieve them within schedule and budget can be traced back to the days when its leaders began to lose confidence in soliciting support for larger programs. During the early 1970s, NASA had several go-arounds with Congress in obtaining approval for the space shuttle. Overruns in the cost estimates and an uncertain political climate forced NASA to retrench until it ended up with a compromise design that tried to be all things to all people. When the President's Office of Management and Budget mandated a ceiling of $5.15 billion (1972 dollars) on the shuttle program, NASA felt compelled to adopt a new strategy sometimes called "success-oriented management." There was no room for mistakes; everything had to go perfectly.

It didn't. The ceramic heat-resistant tiles that were to protect the shuttle from the searing reentry to Earth's atmosphere from space were falling off by the thousands in tests. Technicians had to glue the tiles back on, one at a time. The cost of a single shuttle launch escalated from NASA's original estimate of $10.5 million (1972 dollars) to 1987 cost of well over $100 million, a dramatic overrun even considering inflation.

The claimed savings of the shuttle over expendable launch vehicles became obliterated by the escalating operations costs. A "marching army" of several thousand NASA employees and contractors, who tend to control each shuttle mission, continue to use up resources. The NASA centers bicker over programs, don't often communicate with one another, and create unnecessary duplications. This bureaucratic beast seems to be an unwelcome legacy of Apollo. Meanwhile, many of the best top management people have left, and leaders and policies come and go through a revolving door, sometimes more than once. And then came *Challenger*, history for all to see.

Even some of the optimists are glum. The economist Klaus Heiss, who had helped NASA develop its case for the shuttle, recently said: "I'm much more pessimistic about American interests in space than ten years ago. Europe and Japan are in better shape."* Lamenting that too much of NASA's budget is tied up in operations rather than science or advanced concepts, Heiss concluded that "a lot of NASA is burned out."

The space station is NASA's newest *raison d'être*. But it has many of the same problems the shuttle had when it was first proposed. Once again, NASA is "betting the company" on a single massive effort whose intrinsic value represents a capability rather than a public-arousing mission like Apollo and Mars 1999. And also like the shuttle, the station seems to represent many things to many people but may turn into a rather weak compromise between conflicting interests. Its monolithic construction and budget demands may preempt or delay the R&D needed to create Mars 1999; for example, the Earth orbital variable gravity experiment, precursor to the tethered modules of the *Olive Branch* in the Mars 1999 scenarios.

The space station is a single structure designed to be a microgravity materials processing laboratory, biological research center, astrophysical and Earth surface observatory, and orbital gas station for refueling rockets that would carry crews and equipment beyond. But we can see conflicts arising out of the Erector set approach. For example, what are the effects of moving people and materials on the pointing accuracy of sensors or on the materials experiments? What about possible contamination imposed by the impingement of rocket propellant plumes on the station? Some engineers have suggested physically separating the modules according to function, but then NASA would risk having piece by piece cut away by the Administration or Congress until it had nothing.

"The station serves no great goal," wrote the editors of the *New York Times* on February 12, 1987, "just a multitude of minor missions to muster the widest support from all possible users. . . . Scrapping the white elephant space station would free funds for vigorous exploration of the universe." They proposed the trip to Mars.

In spite of my earlier optimism about the space station, I am now

---

*Reported by Wayne Biddle in *Discover*, January 1987.

concerned about the station consuming most of NASA's resources and per-petuating a bureaucratic juggernaut that is too focused on operational infra-structure and not enough on advanced planning, research and develop-ment, hard science . . . or Mars 1999.

This concern is reinforced by reports, as of this writing, that the space station could be delayed up to two years and cost $13 to $14.5 billion, not the $8 billion originally estimated. In addition, microgravity research ex-perts have warned NASA that materials processing efforts in the United States will need some beefing up, or we will be well behind Japan, Europe, and the Soviet Union. A study prepared by a committee chaired by astro-naut Bonnie Dunbar stated that the U.S. risks becoming "the landlords of the space station, not the tenants."

Timely launch of the unmanned Mars Observer mission would also be important to the Mars 1999 program. NASA has proposed postponing the launch from 1990 to 1992 because of a crowded shuttle launch schedule. Ironically, the delay would cost NASA an additional $100 million because of the schedule slippage.

"NASA today seems moribund," wrote Wayne Biddle in the January 1987 issue of *Discover* magazine. "The space program lacks realistic long-term goals and the sense of initiative they could reawaken. Worse, it seems to lack the leadership needed to formulate such goals and sell them to those who control the money in Washington or to the American people."

Behind the *Challenger* disaster is a crisis in leadership. In their book *Prescription for Disaster* (Crown Publishers, 1987), Joseph J. Trento, an inves-tigative reporter with Cable News Network, and Susan Trento wrote the shocking story of the lack of leadership underlying the explosion.

The Trentos reveal a series of bizarre behind-the-scenes events in high places in Washington. President Reagan is described as almost "technically ignorant" of the space program, according to former NASA administrator James Beggs. The President's science advisers pursued an apparently Machi-avellian course of opposition to Beggs. Military expert William Graham, inexperienced in civilian space leadership, was appointed deputy adminis-trator (over Beggs's objections). Graham succeeded to the position of admin-istrator when Beggs was indicted for allegedly cheating the government years ago in a contract for a new anti-aircraft cannon while he was a vice president at General Dynamics. All the while a breakdown of communica-tion among NASA leaders and their industrial partners was leading step by step to the *Challenger* tragedy. The Trentos' book is full of pithy quotes such as this one from former presidential science adviser George Keyworth: "Of all the organizations I deal with, I have only seen one that lied, NASA. The reason they lie, of course, is because they are wrapped up in a higher calling. In their eyes, [these] are white lies."

It is interesting but not overly encouraging that former administrator James Fletcher, the man who sold us the shuttle, is once again back at the helm selling the space station.

In their February 12, 1987 lead editorial entitled "The Poverty of NASA's Dreams," the *New York Times* editors wrote: "A country with a $4 trillion gross national product can afford a vigorous space program that meets practical needs and stirs people's sense of adventure. NASA's musty plans and cramped vision lack both utility and imagination. The space agency will spend $9.5 billion for next year's space show. Instead of reaching to stretch man's grasp, NASA engineers plan more plumbing."

What can we do with the bureaucratic beast to make way for Mars 1999? Several proposals have been advanced. One is to turn the operations and transportation infrastructure over to the Air Force. Doing that would shift an already defense-heavy U.S. space emphasis even more to the defense sector. Civilian programs like Mars 1999 or an international space station could slip between the cracks and disappear out of sight. Currently, the Department of Defense spends twice what NASA spends on space.

Another solution would be for NASA to divide into two separate organizations. One would be like the old research and development organization it grew out of. The other would facilitate commercial development of the transportation and facilities infrastructure until launch systems and space stations are commercially owned. Users would include the research and development branch of NASA, the Air Force, and private ventures.

An independent and focused R&D NASA could manage Mars 1999 at an average annual funding level of $2 billion during the 1990s plus another $1 billion per year to rent launch vehicles and space station facilities. This would be one-third of NASA's current budget and the largest component of the budget of the new R&D NASA. We could leave plenty of room for developing new technologies and scientific programs like the kind that rode on the coattails of Apollo.

The transition to a new NASA will be inevitably painful but probably not nearly as painful as the unproductive present. The people most affected would be the federal employees and contractors associated with the launch teams of the Kennedy Space Center in Florida and operations teams at the

MARS 1999 AND FOLLOW-ON OPPORTUNITIES

(IN MOST CASES, DATES CAN BE MOVED ± 1 MONTH)

MISSION NUMBER	TYPE OF MISSION	EARTH DEPARTURE DATE	MARS ARRIVAL DATE	MARS DEPARTURE DATE	EARTH ARRIVAL DATE
1	Mars 1999 Opposition Class (outbound Venus swingby)	January 1998	December 1998	February 1999	November 1999
2	1999 Conjunction Class	December 1998	September 1999	January 2001	September 2001
3	1999 Hyperbolic Mars Flyby	April 1999	August 1999	August 1999	April 2000
4	2001 Opposition Class (inbound Venus swingby)	January 2001	September 2001	November 2001	December 2002
5	2001 Conjunction Class	January 2001	September 2001	February 2003	October 2003
6	2001 Hyperbolic Mars Flyby	March 2001	August 2001	August 2001	May 2002
7	2003 Opposition Class (outbound Venus swingby)	August 2002 ———————————————————————▶			
8	2003 Conjunction Class	February 2003 ———————————————————————▶			

Johnson Space Center in Houston. Opening these capabilities to the private sector to provide the best possible launch and operations services should in the long run provide plenty of jobs. The government could supervise the transition period of a few years to minimize unsettledness. Ultimately NASA, the Air Force, foreign governments, and various commercial concerns, as customers, would impose certain requirements (pounds delivered to a particular orbit, for example), and those requirements would be met by the best available launch system and space facilities, provided by private organizations on a competitive basis.

These structural changes are not essential for Mars 1999, but they would enhance its chance of success. In any case, Mars 1999 needn't make an appreciable dent into NASA's budget, even if NASA continues to run the launch and space station facilities at high cost. As mentioned before, the research, development, and purchase of equipment for Mars 1999 should consume only about $2 billion per year, less than one-fourth of the total current NASA budget.

Research and development at the existing NASA centers would prosper under a Mars 1999 program. An example of how the new resources could be divided might look like this: The Marshall Space Flight Center in Huntsville, Alabama, would provide interplanetary spacecraft components such as the modules, aeroshields, the reconfigured external tank, and perhaps the PhD lab. The Johnson Space Center could be responsible for the Mars Excursion Module, crew training, and operations. The Jet Propulsion Laboratory in Pasadena, California, would handle some of the unmanned elements: the Mars surface rovers, balloons, orbiting observers, and some of the scientific experiments. The Langley Research Center in Virginia could provide the PhD processing facility. The Lewis Research Center in Cleveland could supply the solar furnace and power systems. The Ames Research Center in California could supervise life sciences, including variable gravity research, crew accommodations, and physiological and psychological factors. And the Goddard Space Flight Center outside Washington, D.C., could handle the orbital mechanics, communications, and some of the science experiments. Aerospace contractors would also thrive. There is something in this for everybody.

The new R&D NASA would need to follow a charter that ensures continuity through the years. We cannot repeat the patterns of the past, where projects are arbitrarily canceled or postponed because of shortcomings in the transportation system or the encroachment of other programs or policies. Likewise, the program will need to be more immune to the annual budgeting cycle in the Administration and Congress. We will need to build

MARS 1999 AND FOLLOW-ON MISSION OPPORTUNITIES TO PhD AND MARS

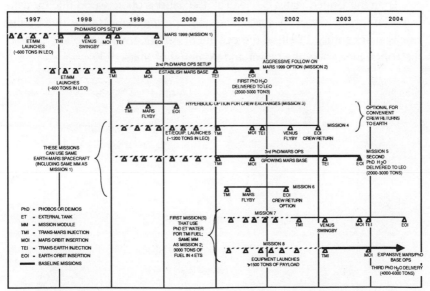

long-term commitments into the system so Mars 1999 and its spinoffs and successors can move ahead. We must allow for some waste and dead ends in the process. Trial and error are a part of what research is all about. We must give ourselves the right to be wrong, to learn from it, and to grow wiser and more competent.

Many of these suggestions may sound like a Pollyanna pipe dream to the hard-nosed, "realistic" policy analysts in Washington. Yet, as the Apollo precedent has clearly shown, great achievements *are* possible with skilled, enthusiastic leadership. The payoff will be enormous—in money, productivity, leadership, peace, and the human spirit.

To run the new NASA, we will need new managers, individuals who can thrive in the context of an aggressive, goal-oriented research and development program like Mars 1999. These new individuals will come from both inside and outside the space program. Many of them will be younger people with enthusiasm, optimism, and vision. When all this happens, and it *can* happen, the technical management of Mars 1999 will be a snap.

Ultimately, there is only one person who can set the wheels of needed change in motion—the President of the United States. He or she can set the goal and reorganize NASA to make it happen. The investment in Mars 1999 would consume only one part in 500 of the federal budget—very small, by weapons standards.

Then the United States can lead all the nations of the world into humanity's triumphant debut in space and the coming renaissance during the twenty-first century. I believe the people of America will be solidly behind that, once they know the power and potential of Mars 1999.

Scenario

2020 Hindsight

• *JANUARY 2001—The* Enterprise *crew, having spent 16 months on Mars and Phobos, launches Earthward with a payload of 2,500 metric tons of Phobos water. The* Olive Branch *with its crew escapes Earth orbit for its second (and the United States' third) journey to Mars. • SEPTEMBER 2001—The* Enterprise *arrives at Earth with a big haul of water; the* Olive Branch *arrives at Mars and Phobos to double fuel-processing capacity. • FEBRUARY 2003—The* Olive Branch *leaves Mars with a second load of Phobos water, and the* Enterprise *is again launched toward Mars on the fourth U.S. manned mission. • OCTOBER 2003—The* Olive Branch *arrives at Earth with a second big haul of water; the* Enterprise *arrives at Mars. • 2005—A total of 10,000 metric tons of Phobos water delivered to Earth orbit in four missions supplies enough fuel to launch all the elements of permanent lunar and Mars bases, which are established later that year. • 2010— Space colonies and settlements begin to spread across the solar system; energy to Earth is supplied by solar power satellites; asteroids are mined for metals. • 2020—We have established world government, peace, telepathic communication with one another and with extraterrestrial beings; we are in a new age.*

The year is 2020 and I am eighty years old, writing this retrospective spanning three decades. We are a species in transition to a new age, a renaissance. More than any other single event, the catalyst was Mars 1999.

Four hundred years ago a small number of courageous Pilgrims landed on Plymouth Rock. Many of them perished, but the survivors had the will and spiritual motivation to settle new lands with new ideas.

We are now witnessing similar events. Lunar colonies, Mars colonies, and space colonies are growing fast. A new world order is emerging with space as a focus. Freeman Dyson had predicted that the financial investment, as a percentage of gross national product, and personal risk would be far less for the space settlers homesteading the asteroids than it had been for the Pilgrims. He turned out to be right.

In the year 2020 astronauts uncover fossils at the Valles Marineris base. *Painting by Pat Rawlings, Eagle Engineering.*

We see world government emerging, as envisioned in the 1980s by Joanne Gabrynowicz; space was the catalyst. The ongoing constitutional conventions in Geneva, which I have been delighted to participate in, are putting the final touches on a solid political foundation. What had been at first an international cooperative exploration of space, and then Mars 1999, turned into explosively growing economic partnerships centered on PhD propellant processing, asteroid mining, solar power satellites, food production in space, and space tourism. This multitrillion-dollar enterprise created a World Common Market. For a while many of the Third World nations objected, wanting full representation. But when they learned there was plenty to go around, they jumped in. Also, the concepts of a world senate and house of representatives provided a good compromise to the lesser developed nations. A political union of the world was the logical next step.

How did all this come about during the twenty years after our great adventure to Mars? The tangible benefits began rolling in during 2001. The *Enterprise* returned from Mars in September of that year, flying alongside two external tanks with payloads of 2,500 metric tons of PhD water. The *Ironhorse* was back in Earth orbit, the remarkable vehicle that served as a

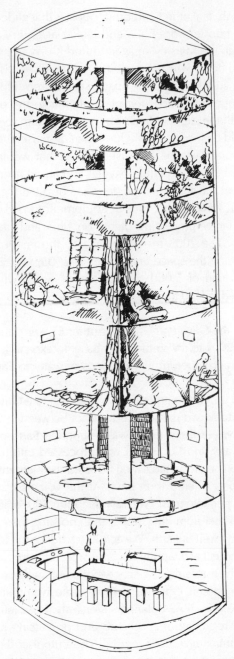

Hab module of the interplanetary spacecraft. *Drawing courtesy SCI-ARC.*

booster from Earth, tanker in Earth orbit, booster that glided us toward Mars, tanker on PhD, booster from PhD, and tanker again in Earth orbit. Later it would go through the Mars cycle again. I find it very impressive that an old tank that used to be thrown away into the Indian Ocean is now so versatile.

The water hauled back to the space station from PhD poured into an Earth orbital propellant processing plant similar to the one on PhD. This plant began to fill the external tanks for the next return to Mars, in February 2003. This would be the second flight of the *Enterprise*, the *Ironhorse*, and a second ET. It would also be the first mission that would use PhD water processed in Earth orbit into liquid oxygen and liquid hydrogen. During the 2003 launch from the space station, four ETs boosted the *Enterprise* plus a payload of about 1,500 metric tons of equipment toward Mars. The equipment included larger processing plants on PhD and the first permanent Mars settlement. The 2003 mission was also the first to yield huge dividends in expanding the space enterprise. In 2005, four ETs returned from PhD with a payload of 5,000 metric tons of water, which supplied the propellant needed to land the first lunar settlement and to supply it with life-support water.

Meanwhile, the *Olive Branch* and another ET were launched from Earth orbit in January 2001 to revisit the Mars base, to expand it, to process more PhD propellant, and to return two ET loads of water to Earth orbit in October 2003.

All summed up, two interplanetary spacecraft (the *Olive Branch* and the *Enterprise*) each took two round trips to Mars between 1998 and 2005. Six external tanks, two of them used twice, supplied Earth orbit with 10,000 metric tons of water, most of which was processed into liquid oxygen and liquid hydrogen. The propellants then boosted the elements of a permanent Mars base and permanent Moon base. At the going Earth-to-space launch cost of $6,000 per kilogram, we saved about $60 billion by obtaining propellants and water from PhD, rather than hoisting them from the bottom of the deep gravity well of Earth. We actually paid for the four missions! The lunar base and Mars base rapidly became self-sufficient, and new missions set out to mine the asteroids. The space explosion had begun, with 2005 as the year of the economic crossover toward explosive profit.

During those years NASA planners elegantly timed launches from Earth to orbit and the recycling of spacecraft and processing rates to minimize investments and maximize their returns. The output of the PhD enterprise

Lab module of the interplanetary spacecraft. *Drawing courtesy SCI-ARC.*

doubled every two years, and the asteroid enterprise promised to increase yields at an even greater rate.

By 2010, we had space colonies and settlements that began to spread across the solar system. Space soon became the prime source of energy to Earth. Solar power stations placed on the Moon and in synchronous orbits around Earth collected the Sun's energy, converted the energy to microwave beams, and transmitted them to receiving antennas on Earth. Nations collaborated to eliminate polluting nuclear and fossil fuel power plants.

Mining operations on the asteroids began to supply Earth with billions of dollars' worth of platinum, iridium, palladium, germanium, and gallium. And huge space farms fabricated from the asteroids began to drop off food to needy nations as part of a world hunger relief project.

Space tourism has become the latest boom in the economy. Last year alone Society Expeditions transported 4,000 people to the lunar settlement hotel, and more than a million people have now taken the trip into Earth orbit. With the new mass drivers and solar sail propulsion systems being built, there is talk of opening up Mars to tourism. The Research and Development NASA just authorized the construction of the first starships that would be fueled by fusion engines and lasers. About 200 pilgrims will be picked to take the trip in 2022.

The increasing number of contacts with extraterrestrials and the spread of a larger spiritual awareness across the solar system certainly has made life more interesting in the twenty-first century.

Marla Lee and I were married in 2000. We live in Hawaii, and although I am octogenarian and she is in her sixties, we both feel younger than ever. Our lives are creative and settled.

As we now gaze toward the sky at night, we can see bright stars along an equatorial band in the sky. These are solar power satellites, a new source of electricity for Earth. Beyond lies the quarter Moon. My son is up there now, working as an artist for the University of Hawaii, which is completing an expansion of its radio telescope array on the lunar backside. Free of radio noise from Earth, this facility has already picked up signals from several extraterrestrial civilizations.

And then there is the red planet Mars to the east, winking at us from between the satellites. My daughter, son-in-law, and their family became colonists on Mars. Their community in the Cydonia pyramid complex is an extraordinary experiment in living. The inhabitants communicate telepathically, and their consciousnesses can travel out of their bodies at any time.

They are learning the mysteries of the universe from their extraterrestrial partners and from inside themselves. The information coming through is extraordinary. Now that Earth has calmed down from its twentieth-century crises, many of us here are also becoming privy to realms beyond our wildest imaginations.

My daughter talks with Marla and me telepathically and instantaneously. The speed of light is no longer a limit. We can now see her in our minds' eyes sitting on a reclining couch inside a huge plastic bubble on Mars. Lush green plants surround her. Outside the bubble we can see a dust storm raging with hundred mile-per-hour winds.

"Dad," she said to me, "when you were here, you said you felt strangely connected with everything and everyone, but you haven't said much about it since then. I feel that way all the time!"

"That's good, my dear, because I sometimes forget. I forget that nothing is impossible except what my mind tells me."

12

Living the Vision

Painting the scenarios of the odd-numbered chapters of this book was a challenge for me.

It is often said that the truth is stranger than fiction. Yet the science fiction we read usually involves far-future fantasies that may or may not ever really matter to us except as an escape from today; realistic projections of technology are unimportant. The real challenge to me is to bridge the gap with technically plausible scenarios that are neither outrageous nor naive. We know that the emerging truth will blow away our most "plausible" predictions. It always does.

Times were simpler during the 1940s and 1950s when Wernher von Braun, Willy Ley, Arthur C. Clarke, and others were describing their scenarios. Now we seem to be trapped within a Tower of Babel, with little incentive to create these plausible scenarios that tell the story of what we can do, if we are inspired enough to go for it.

Policy requires goals, but we have no real goals in space. So, by default, space policy has become a chaotic process of bureaucratic preservation. What we need are clearly expressed goals with a coherent set of objectives that capture the public's imagination and free up resources for advancing the health, wealth, and happiness of humans on Earth.

How can we resolve our current dilemma? I have come to the realization, after many decades of personal striving, that each and every one of us must, to the degree that we can, *live* our visions. Meanwhile, we need to be tolerant of the world as it is, realizing we are here to learn and to grow.

Let me give an example. Years ago, I had some visions of asteroid mining that became well reconciled with my scientific mind, and I wanted to publish the ideas in the professional literature. Even with the support of a few colleagues such as Gerard O'Neill, it took two years for my words to reach print in the journal *Science*. This happened ten years ago. Now asteroid mining is a common topic of discussion among space futurists and is included in the plan of the National Commission on Space.

The President's National Commission on Space also had a vision. Theirs involved a number of steps including a lunar base, the human trip to Mars, cycling space stations between Earth, the Moon, and the planets, and eventually materials processing at Phobos, Deimos, the Moon, the asteroids, and Mars. Their vision is very appealing and represents a series of goals and incentives beyond NASA's purview. Unfortunately, the report came out shortly after the *Challenger* disaster. Revelation of their findings was eclipsed by the report of the Rogers Commission on *Challenger*, a much more newsworthy happening.

But the commission's report does not contain the one ingredient I believe essential: one dramatic event that would stir the public interest and trigger the renaissance of opportunities we are yearning for. Their focus is rather on a series of practical steps that allow the infrastructure to grow gradually—for example, a human return to the Moon in 2005 and the first manned trip to Mars in 2015. But as we have seen in this book, all these things can happen sooner and more economically if we go to PhD first.

Are we ready to achieve our destiny in space? Were the *Challenger's* seven final crew members tragic victims of human ineptitude, or were they leading us all into an experience of a greater truth? Can it be said that huge multilateral investments in the arms race, Star Wars, and other pursuits worldwide are a collective illusion, a nightmare scenario of a game with very high stakes? We need more than that now.

I have come to the punch line of this book. To a degree that may be unconventional for a scientist, I have expressed my quest, a story of how we here on Earth can transcend our gravity-bound experiences and create a grand human debut in space and a renaissance on Earth. The quest is, how can we go assuredly into space and thus create a renaissance for humankind? Mars 1999 is my expression to the credible engineers, scientists, and leaders of the world. With little cost and effort and with lots of inspiration, it can be done. Believe me, it is very powerful, once you catch the updraft.

More powerful still is the arena of the spirit. By "spirit" I am looking beyond religion or any other organized body of beliefs and practices. By "spirit," I mean the inspiration to fly away from the quagmire of gravity and limitation toward greater heights, both literally and figuratively, and toward greater truths about the human potential. Mars 1999 is a physical metaphor for human transcendance. The historical record is full of great scientists who would understand and confirm the essence of this metaphor. It is nothing new. It is always available.

NASA planners or Washington space pundits might now bet a lot of money that Mars 1999 will not happen. They would dismiss it as an overly optimistic and unrealistic projection of some space scientist's wish list. They would say it's too much too soon, that we can't do it because it violates current paradigms of infrastructure planning. If we humans opt for such a limited view, then that is what we will get. It will become a self-fulfilling prophecy, to no one's benefit or credit.

And here we come to my opinion, my personal point of view: Mars 1999, in and of itself, is probably too conservative. An old order is dying and a new order is emerging. The knowledge of the ages and the beauty of this universe are *inside* of us, far more than all around us, and are being unlocked to many of us at an accelerating rate. By simple virtue of my training as a physicist, I am confronted always by the limits of our senses, our minds, and our emotions. Mars 1999 is an attempt to carry the rules and assumptions of an old order to a reasonable next step. Meanwhile, more radical worldwide revelations, actions, and breakthroughs that are likely to take place will give Mars 1999 a bolder context and make it even easier to implement.

My projections merely point out possibilities that lie within our limited, linear-conditioned minds. Mocking up an expansive, positive future beyond the purview of Mars 1999 is where you, the reader, come in. I invite you to probe the limits of your imagination, to live your vision. If enough of us do that, and if the vision is shared by enough people, we shall not only live it; it will become physical reality.

A universe unfolds around us and inside us. We are on the threshold of a renaissance that can be triggered by our full-fledged debut in space. We are in free fall toward a future beyond our wildest imaginings. Mars 1999 is a symbol of the constriction of the hourglass, and so it looks to us like time is running out for us on Earth. We can react out of fear, or we can choose to respond to the opportunity and drop our illusory considerations. We can venture through that opening to a new age where time opens up before us and the universe is ours to explore. You decide, fellow voyager; then live your vision and take your rightful place. You won't regret it. I'll see you there.

Epilogue

Two of the characters of the fictional part of this book are not fictional: Sevastyanov and myself, the two individuals who, in the story, land on Mars. In real life, Sevastyanov, a former Soviet cosmonaut, and I have met three times. On all occasions, we have had a connection of the heart even though we didn't know each other's languages. I last saw him at the U.S. embassy in Prague, Czechoslovakia, in September 1977, exactly ten years before the publication of this book. The occasion was a reception for space scientists and engineers who had come from all over the world to attend a conference. We spotted each other through the crowd, ran across the room, and hugged warmly and enthusiastically. It was a joyous moment, one I hope will be renewed soon.

Letter to President Reagan

April 9, 1986

President Ronald Reagan
The White House
Washington, D.C.

Dear Mr. President:

This is my first letter to you. I consider it so important I am hoping your aides will give you a chance to read it. As a former NASA astronaut, U.C. Berkeley astronomer, and now chairman of the National Space Council, I come from a credible and knowledgeable position on what is possible in long term space policy.

By taking an initiative only you have the ability to perform, I believe you have a unique opportunity to become a statesman who will be remembered in history as being the president who turned the arms race into peaceful purposes, who opened the international community to cooperation in space and on the Earth, and who created an infrastructure that would lead to nothing less than a human renaissance in space. I believe what you could do over the next year or two would be historically of much more value and impact even than John F. Kennedy's dramatic Apollo initiative.

What I am proposing, Mr. President, is that you set as a goal a joint U.S.–Soviet mission to Mars under a set of circumstances that would safeguard national security and permit the United States to take a leadership role that would yield extraordinary scientific, technological, political, economic, and spiritual dividends.

As you know, Mr. Gorbachev's proposal for phased reduction of nuclear weapons leading to complete nuclear disarmament by 2000 is attractive but poses some practical difficulties, not the least of which are what credit you would be given in formulating the proposal itself, and what work could be performed by the space military infrastructure—particularly the Strategic Defense Initiative (SDI)—if that project were to be curtailed or cancelled, as it would need to be in a nuclear disarmament scenario.

Your proposing the joint mission to Mars would answer both difficulties as part of the disarmament package. But to safeguard U.S. (and Soviet) interests I would suggest that the missions be performed in parallel, separate spacecraft flying close to each other in the event one nation were to pull out. The two craft would be designed to dock with each other for crew exchanges during the tedious 22-month mission and could also perform as backup emergency rescue.

The concept I suggest is to first land on one of the moons of Mars, Phobos or Deimos, or PhD (the Soviets are going to Phobos in 1988, unmanned, presumably in preparation for manned missions there during the 1990s). Because they have little gravity, it is two to three times less expensive to go to PhD than directly to the Martian surface. In fact Phobos and Deimos, in terms of fuel cost, are even more accessible to us than our own moon!

Then an international scientific base would be set up on PhD that would remotely control unmanned rovers and sample returns from the Martian surface (the Earth is too far for controlling rovers because of minutes' light time distance). After samples are screened for possible contamination on PhD and if all systems are safe, one astronaut and one cosmonaut would board a small sortie module and take the journey to the Martian surface, plant flags, collect samples, and ascend to PhD, all within a few hours. This would be the only part of the mission that would require U.S.–Soviet joint participation.

An excellent launch opportunity will occur in 1998, with a Venus outbound swingby, two months spent near Mars, and a return in November 1999–in time to celebrate complete nuclear disarmament and the turning of the millennium!

The cost of the mission to each nation would be less than $25 billion, and would make use of technology both countries are developing–principally space station modules and orbital transfer vehicles with aerobraking capability. The NASA budget would not need to increase and European/Japanese participation would reduce costs further.

Perhaps most importantly, the PhD mission will not be an economic dead end like Apollo. The likely presence of water on these moons, awaiting confirmation in 1989 by the Soviets, means that billions of dollars can be saved by refuelling and restocking life support systems. PhD water could be delivered to such diverse places as the lunar surface and geosynchronous orbit at far less cost than launching the fuel from the Earth's deep gravity well. Soon thereafter, the self-sufficient space infrastructure recommended by your National Commission on Space and the ideas of Gerry O'Neill and others on space colonies could become realities.

The scientific yield of the PhD mission would be far greater than either the direct manned mission to the Martian surface or an unmanned sample return mission. Roving vehicles could traverse hundreds of kilometers at disparate sites, including those that show some tantalizing suggestions of life having been there.

Mr. President, these ideas represent the results of engineering studies performed by leading scientists in the U.S. and abroad. They are an important part of the recent report of the National Commission on Space. The technology and timetable are appropriate (12 years instead of JFK's 8 years in Apollo). They are supported by a large segment of the public, based on opinion polls. The concept represents the interests and desires of an unprecedented confluence of individuals and groups–

manned space flight, unmanned space flight, arms control, unrisky international cooperation, the economic development of space, and scientific discovery. It is the fulfillment of the "bold" goal you see for humanity as we leave the cradle of Earth into the 21st Century.

With NASA at a bureaucratic crossroads, you have a wonderful opportunity to redirect its energies and also to utilize our military space brainpower. During this period of reassessment we must recall the National Aeronautics and Space Act of 1958 "to explore space for peaceful purposes for the benefit of all mankind and to cooperate with other nations or groups of nations in pursuit of the Act's objectives." The Act was recently affirmed by your signing in 1984 of joint U.S. House-Senate Resolution 236 which provides for U.S.-Soviet cooperation in space as an alternative to an "arms race, which is in the interest of no one."

Mr. President, it is time for us to consider options that transcend the near-term interests of groups and nations, and you are in a unique position to lead in that transformation. You would have a lot of support in it. While we can all acknowledge the Soviets' abysmal record of civil rights and propaganda, we cannot ignore active discussion of positive scenarios of arms control and increasing cooperation. The opportunity is yours to act on this, to fulfill our most heartfelt dreams for exploration, for peace, for a renaissance.

Enclosed are relevant preprints and reprints; I'm sure you'll be increasingly apprised of these ideas through the public press and scientific meetings such as the forthcoming AAAS, but the timing is right for you to take leadership now.

I would appreciate your reply and the opportunity to present more information about these concepts.

Sincerely,

Brian O'Leary, Ph.D.
Apollo Astronaut and
Chairman, National Space Council

Glossary

AEROBRAKING. The use of a planetary atmosphere to "kill off" the excess incoming velocity of a spacecraft.

AEROCAPTURE. An aerobraking maneuver that results in the spacecraft being captured by the gravitational field of a planet.

AEROSHIELD. A heat-resistant shield that protects a spacecraft from the searing heat of aerobraking.

APHELION. The point in the orbit of a planet that is farthest from the Sun.

ASTEROIDS. Small planets that orbit the Sun, mostly between the orbits of Mars and Jupiter; they range up to 1,000 kilometers in size.

CHEMICALLY BOUND OXYGEN. Oxygen in the soil of a planet, satellite, or asteroid that is combined in oxide form with a metal or silicon.

"CONJUNCTION CLASS" MISSION. A three-year round-trip mission from Earth to Mars that requires the least amount of propulsion; opportunities occur every two years.

CYDONIA. A region of Mars around latitude 40 degrees north, characterized by plains, mesas, and the intriguing "face on Mars."

DEEP GRAVITY WELL. A planet with a strong gravitational field requiring large amounts of propellant to escape (Earth, for example).

DEIMOS. The outer moon of Mars, about ten kilometers across.

DELTA-V. A measure of the velocity increment (or propellant) required to transfer a spacecraft from one point or velocity to another.

DEPLOY. To place equipment into working order at its destination.

ELECTROMAGNETIC CATAPULTS. Devices that use electricity and magnets to move payloads into space.

ET. The external tank of the space shuttle; it could also be used as an interplanetary rocket and storage tank for propellants.

EVA. Extravehicular activity, or space walk.

FLYBY. A spacecraft or mission that involves swinging past a planet without stopping there or orbiting it.

G. A measure of the acceleration of gravity on a planet or on a spinning or aerobraking spacecraft. One g is the acceleration of gravity on Earth.

GEOSYNCHRONOUS. An equatorial orbit whose period matches the rate of spin of a planet so that a satellite as seen from the planet will remain at the same point in the sky.

GEO OPS. Operations in geosynchronous (24-hour) orbits.

G-LOADING. The amount of gravitational acceleration exerted on a spacecraft and its inhabitants.

HAB MODULE. The habitability module of a space station or interplanetary spacecraft.

HUB MODULE. The module inhabiting the axis of spin of an interplanetary spacecraft.

HYPERBOLIC. An incoming or outgoing velocity of a spacecraft in excess of the gravitational escape velocity.

INFRASTRUCTURE. In the context of a space program, this is a series of missions and systems that are interrelated so that economic or scientific growth and synergy occur.

LAUNCH WINDOWS. Opportunities to launch spacecraft from one planet to another; for a given class of mission, launch windows between Earth and Mars occur every two years and last for about one month.

LEO. Low Earth orbit. This is the class of orbits that applies to spacecraft up to a few hundred miles above Earth's surface. The space shuttle, space station, and assembly and launch site for the Mars craft will be in LEO. In terms of energy, LEO is halfway out of Earth's strong gravity field.

LOX/LH$_2$. Liquid oxygen and liquid hydrogen; the fuels that can be processed from water launched from the Earth or extracted from PhD.

MASS DRIVERS. Electromagnetic catapults that can be used to transfer materials in space or off a planetary surface.

MEM. Mars Excursion Module; the sortie vehicle that will land people on Mars.

MRV. Multiple Reentry Vehicles, used in nuclear warheads to deceive would-be interceptions.

OLYMPUS MONS. The largest mountain on Mars, an extinct volcano three times the height of Mt. Everest and the size of the state of Pennsylvania.

OTV. Orbital Transfer Vehicle; a rocket that moves payloads from one point or orbit in space to another.

PAYLOAD. The cargo and personnel aboard a spacecraft.

PERIHELION. The point in the orbit of a planet that is nearest to the Sun.

PhD. An abbreviation for the two moons of Mars, Phobos and Deimos.

PHOBOS. The inner moon of Mars, about twenty kilometers across.

RCS. The reaction control system of a spacecraft that maintains its attitude and keeps it from tumbling.

REM. A unit of measurement of the human exposure or dosage of radiation from cosmic rays, solar flares, and nuclear power sources.

RETRO FIRE. A rocket blast that removes a spacecraft from orbit and allows it to descend to a planetary surface.

RETROROCKET. The engines and propellants that are fired to allow a spacecraft to descend to a planetary surface.

SDI. The Strategic Defense Initiative, or "Star Wars," a space defense system proposed by President Reagan, designed to contain an offensive nuclear attack.

SETI. The Search for Extraterrestrial Intelligence, a scientific project that involves principally the use of radio telescopes to "listen" to signals coming from possible extraterrestrial civilizations.

SOLAR FLARES. Energetic particles coming from the Sun, associated with sunspots.

SOLAR FURNACE. A mirror that concentrates sunlight onto materials and heats them up.

SPECTROSCOPY. The science of dividing light into its component wavelengths or colors, to determine chemical composition.

TELEMETRY. Radio signals that transfer data between various points in space or between planets.

TELEOPERATE. The use of telemetry to remotely operate a machine, such as a roving vehicle on Mars.

TETHER. A cable joining two or more spinning spacecraft.

THARSIS RIDGE. A highland region on Mars to the west of Valles Marineris, containing giant extinct volcanoes.

THROTTLE BURN. The act of varying the intensity of rocket blasts to modify the velocity of a spacecraft.

TRAJECTORY. The path of a spacecraft acting under the gravitational influence of a planet.

VALLES MARINERIS. The giant Grand Canyon of Mars; it is five kilometers deep, and as long as the United States is wide.

"WATER OF HYDRATION" ABSORPTION BAND. A spectral feature indicating the presence of water bound in the soil of a planet, satellite, or asteroid.

Index